Future Histories

To augment this reality, download the Moving Marvels app.© Moving Marvels 2019

Future Histories

What Ada Lovelace, Tom Paine, and the Paris Commune Can Teach Us about Digital Technology

Lizzie O'Shea

London • New York

For Janet Elizabeth O'Shea

First published by Verso 2019

1 3 5 7 9 10 8 6 4 2

Verso
UK: 6 Meard Street, London W1F 0EG
US: 20 Jay Street, Suite 1010, Brooklyn, NY 11201

versobooks.com

Verso is the imprint of New Left Books

ISBN-13: 978-1-78873-430-1
ISBN-13: 978-1-78873-432-5 (UK EBK)
ISBN-13: 978-1-78873-433-2 (US EBK)

British Library Cataloguing in Publication Data
A catalogue record for this book is available from the British Library

Library of Congress Cataloging-in-Publication Data

Names: O'Shea, Lizzie, author.
Title: Future histories : what Ada Lovelace, Tom Paine, and the Paris Commune can teach us about digital technology / by Lizzie O'Shea.
Description: London ; Broklyn, NY : Verso, 2019. | Includes bibliographical references and index.
Identifiers: LCCN 2018060356| ISBN 9781788734301 (hardback) | ISBN 9781788734325 (UK EBK) | ISBN 9781788734332 (US EBK)
Subjects: LCSH: Computer science—Philosophy. | Automation—Social aspects. | Digital communications—Political aspects. | Technology—History.
Classification: LCC QA76.167 .O84 2019 | DDC 004—dc23
LC record available at https://lccn.loc.gov/2018060356

Typeset in Fournier by MJ & N Gavan, Truro, Cornwall
Printed and bound by CPI Group (UK) Ltd, Croydon CR0 4YY

Contents

Acknowledgment of Country

I wish to acknowledge the traditional owners of the land on which this book was written. I pay my respects to the elders of this country and to the custodians of the land today. I offer them solidarity. In Australia and in the United States, that land was stolen, and sovereignty was never ceded. I am grateful for the opportunity to live on this country. It always was and always will be Aboriginal land.

1

We Need a Usable Past for a Democratic Future

A Spanish Prince's Automaton and an American Novelist's Living History

Don Carlos was seventeen years old in April 1562 when he fell down the stairs and hit his head. He was the heir to the Spanish throne, studying at the university town in Alcalá de Henares. Depending on who you ask, he was either something of a lush lothario or an inbred oddball (his parents were half-siblings). One observer noted his "violent nature, his intemperate speech and his gluttony." But the reports also indicate that he was well liked by the Spanish people, as a teenager at least. His whole life reads like the plot of a modern gothic fantasy television series: allegations of treachery, leading to solitary confinement at the hands of his father, an episode of bingeing and purging, and ultimately death, possibly by poisoning. His life was later the subject of Giuseppe Verdi's great opera *Don Carlos*.

But all that drama was yet to come, when, while still a young man, engaged "possibly on an illicit errand," as one scholar politely puts it, he tumbled down a disused flight of stairs and knocked himself out on a closed door.

In these early years of Don Carlos's life, relations with the paterfamilias were still good, and the king was devastated by his eldest son's misfortune. He was bedridden by his head injury. Numerous doctors flocked to his bedside, and Don Carlos was subjected to a variety of barbaric surgical procedures, including a misguided attempt to drill a hole in his skull. He eventually fell into a coma and was expected to die.

The local people were very upset by their prince's malady. In an effort to help, they brought Don Carlos the century-old relics of a former member of the local Franciscan order of friars. Since they wanted this friar to be canonized, his body was presented to the prince in hopes of a miracle. The "desiccated corpse" was brought to the prince's bedside, where, unable to open his eyes, he reached out to touch it, then drew his hands across his feverish face.

Suddenly Don Carlos made a remarkable recovery. By the following month, he was back to his usual self. His doctors were stunned. Reflecting on the brutality of his later life, it is unclear if his survival was a blessing or a curse. In any event, the desiccated friar was made a saint.

The prince's own explanation for his recovery was that the figure of a man, "dressed in a Franciscan habit and carrying a small wooden cross," came to his sickroom and assured him that he would recover. This, scholars suggest, was the inspiration for what must be one of the world's most fascinating objects: an early automaton of a friar.

Today the automaton is held in the Smithsonian. In a history that reads more like a detective story than an academic article, this minor miracle of engineering is described by professor Elizabeth King in the following terms:

> made of wood and iron, 15 inches in height. Driven by a key-wound spring, the monk walks in a square, striking his chest with his right arm, raising and lowering a small wooden cross and rosary in his left hand, turning and nodding his head, rolling his eyes, and mouthing silent obsequies. From time to time, he brings the cross to his lips and kisses it. After over 400 years, he remains in good working order.

The workings of the friar are concealed beneath his cloak, fashioned from wood, but the inner levers and cogs are beautifully made, though they were designed to be seen by no one but the maker. This shell gives the figure an air of ghostly mystery, inspiring fear and reverence in all who witness him move about without visible assistance, as if by magic.

No one really knows where the friar came from. King's thesis is that the creator was Juanelo Turriano, an engineer who worked for King Philip. A prodigy from humble origins, he became a distinguished maker of astronomical clocks and other similar instruments, and even designed a system of waterworks for the city of Toledo. After his son's impressive recovery, Turriano could well have been commissioned to build the contraption by King Philip in honor of the Franciscan friar, who was deemed responsible for this miracle.

King ascribes the creation of the automaton friar to what she terms an "ambitious impulse," the ancient and abiding human desire to understand by imitation. She argues that it recalls Descartes's thinking about the connection between body and mind —questioning whether we are driven from without or within. "The automaton forms an important chapter in the histories of philosophy and physiology," writes King, "and, now, the modern histories of computer science and artificial intelligence."

Objects like clocks and automatons are in many ways the predecessors to modern digital technology. You needed to be both an engineer and an artist to build these kinds of machines—technology was often entertaining, inspiring, frightening and useful, all at the same time. In this sense, the path to the modern networked computer was paved with excruciating care and dedication, as well as a little whimsy. It was a journey populated by experimentation with both functional and decorative objects, and those who work with equivalent kinds of advanced technology carry on this tradition today.

Examining this mechanical friar through twenty-first-century eyes, we recognize many themes of our history and our future, our excitement and misgivings about our current relationship with technology. The friar shows how stories from our past can shape our destiny. Our past tells us about our present—how it was just one of many possible futures claimed by those who came before. In this context, both the creation and use of technology express a kind of power relation. King writes about this, in summarizing conversations about the friar with the Smithsonian conservator, W. David Todd:

> Would the measure of the monk's power have come from the sight of a king setting him in motion? But Todd and I agree the power flows in the opposite direction, so that once the tiny man is seen to move independently, the operator's status takes a leap, *he* becomes a kind of god. Either way there is a mutual transfer of authority and magic. Todd, jesting only a little, likens the possession of the monk to owning the pentium chip a couple of years ago. Who commands the highest technology possesses the highest power.

If we accept King's hypothesis, the friar is a product of royal decree and religious fervor, serving as a tribute to divine intervention but made with a very human, highly material skill. Today the leading edges of technological development are occupied by similarly powerful individuals, who use technology to inspire loyalty and also to intimidate. The "magic" of modern technology implies that the trajectory of the digital revolution is objective and unassailable and that the people driving its development are great figures of history. Technological objects, even those that are or seem to be playful or diverting, are designed with a certain purpose in mind, and they can influence us in profound ways.

But Don Carlos's automaton also tells us something about how technology is produced in contemporary society. The friar is a piece of craftsmanship that has lasted four centuries, whereas a comparable artifact today might be built in a Chinese factory, under appalling conditions, complete with planned obsolescence. Such a contrast demonstrates how technology is a field of creativity and skill, especially in its early, innovative stages. But when it is scaled up, it can become an industry of exploitation. The promise of technology has always relied on the meticulous efforts of people like Turriano; yet concealed in many beautiful objects that we see and handle every day is the brutal labor history of places such as Shenzen that testifies to the power of the process of commodification. Having replaced artisanal automatons with mass-produced robots, we start to treat others and feel like robots ourselves. Our current society reveres some kinds of labor and debases others, and the power of technology to improve our world and livelihood is not equally distributed.

The past lives on in memories and stories and in the objects we use and produce. The networked computer represents an exciting opportunity to reshape the world in an image of sustainable prosperity, shared collective wealth, democratized knowledge and respectful social relations. But such a world is only possible if we actively decide to build it. Central to that task is giving ordinary people the power to control how the digital revolution unfolds.

In the huddle of people attending Don Carlos, amid all the hubbub of miracles and reverence, one doctor did claim that his recovery was due to objective factors rather than divine intervention. "The cure was of natural origins," he bravely argued,

> only those [cures] are properly called miracles which are beyond the power of all natural remedies ... People cured by resorting to the remedies of physicians are not said to have been cured by a miracle since the improvement in their health can be traced to those remedies.

This pert remark serves as one doctor's message to the future, to those who would come after him. Seek evidence, speak honestly, he seems to be saying, try to shine a light of truth on the events to which you bear witness with integrity. Cause and effect exist in the real world, and humans can both observe this process and sometimes influence it with their agency. Do not be distracted by religious ardor or royal conceit.

We can still see the glimmers of this light, even four centuries later. Turriano created a marvelous and beautiful object that commemorated the recovery of Don Carlos, and he contributed to our collective technological knowledge, the legacy of which lives on in computing today. But his work was made to pay tribute to divinity rather than stand as a testament to human ingenuity and science. It is not hard to imagine how the formidable skills and creativity on display might be used to tackle some of the problems faced by humanity. But only if we take the power out of the hands of kings.

This is not a book about technology per se, nor is it about history or theory. Rather, it is an attempt to read these things together in fresh and revealing ways. The purpose is not to comprehensively or categorically define the nature of the problems we face in digital society or offer prescriptive solutions; it is to suggest ideas and identify points of conflict. Not to provide definitive or exhaustive histories of certain events or schools of thought but to start a conversation about how certain histories are critical to the task of designing our future. It is written for those who may be knowledgeable about technology but lack an understanding of radical and democratic political traditions, and for those who, while familiar with such theory and practice, are wary of or inexperienced with digital technology. My aim is to find a common language for a more sophisticated discussion about the future of both topics, which must be predicated on an agreed understanding of the past. I mean to anchor the present to the past for a specific purpose: to argue that democratic control of digital technology—building structures that give people more say and control over how digital technology is produced and developed—gives us the best chance of overcoming some of the problems we face today. It is about creating a "usable past" for digital technology, a concept that has its own little history.

Van Wyck Brooks was a writer and critic when American literature came of age in the early twentieth century, a person profoundly committed to literary practice and culture. His voice "exhort[ed] writers to meet their responsibility with courage and dignity—and with pride." This led him, in 1918, to call for the creation of what he called a "usable past." Speaking to his contemporaries in an intelligent and vivid essay, he outlined the need for history that creative minds could draw upon. "The present is a void," he wrote, "and the American writer floats in that void because the past that survives in the common mind of the present is a past without living value."

It is understandable that younger generations are eager to look forward. History can weigh like a millstone; archaic distinctions and practices can drag upon our freedom and agency. But detachment from the past has its own pitfalls. It means that the past that survives is a default genealogy, a mere reflection of the status quo, fixed and irrelevant. It loses its living value, its capacity to help the current generation actively shape a collective sense of self, leaving us isolated without a common sense of purpose or a forum to discuss these ideas. "The grey conventional mind casts its shadow backward," Brooks observed. "But why should not the creative mind dispel that shadow with shafts of light?" For Brooks, the American literary history of the nineteenth century was important to document because it showcased the beauty, daring and distinction of American artists. It was a task to which he devoted years of his life, reading 825 books for his literary history *The Flowering of New England* (1936). This monumental task was part of his aspiration for "cultural centralization"—to create a communal language and bring to life a common culture and identity.

The purpose of a usable past is not simply to be a record of history. Rather, by building a shared appreciation of moments and traditions in collective history, a usable past is a method for creating the world we want to see. It is about "cutting the cloth" of history, as Brooks put it, to suit a particular agenda. It is an argument for what the future could look like, based on what kinds of traditions are worth valuing and which moments are worth remembering.

A century later, Brooks's challenge to the American literary community retains its relevance in the age of digital technology. The digital revolution is creating experiences that are sometimes exciting, often horrifying, and routinely amazing. But present discussions about our digital future seem to float in a void. A whole set of assumptions about the past, static and dry, occupies our consciousness. It is as though digital technology sprang from nothing, invading private spaces and public life like a juggernaut. The merit of organizing our lives around screens is rarely questioned, and we wear objects that endlessly track our movements and sometimes literally get under our skin. A commitment to meritocracy saturates public debates about technology, and freedom is understood in atomized and commodified terms. There is tacit acceptance that governments and corporations will determine the evolution of digital technology. It is also widely accepted that it is easier to imagine the end of the world than the end of capitalism—an assumption that persists even during the most transformative moments in technological development.

Digital technology is treated as a force of nature, without an agenda, inevitable and unstoppable. The past that has survived in the minds of the current generation is one that reflects what has happened rather than what is possible. Society is often treated as an object, which digital technology does things to, rather than a community of people with agency and a collective desire to shape the future. "All our invention and progress seem to result in endowing material forces with intellectual life, and in stultifying human life into a material force," declared Karl Marx. Nowhere in our current society is this observation more relevant than our personal and political engagement with digital technology.

For Brooks, the starting point was to ask: "*What is important for us?*" His focus was building a sense of identity among the American literary community, to find what was distinctive and valuable about the American voice. His starting point still has value. In the context of the digital age, what is important for us? What is distinctive and worthwhile about digital technology, and how can it be used to enable humanity to flourish?

Another world is possible, where society is collective and humans have agency over their digital futures. But to get there we need to create a past with living value.

In part, the motivation for this book comes from observing the ahistorical nature of discussions about technology. This has, at best, led to a benign yet thoughtless form of technological optimism.

"When you give everyone a voice and give people power, the system usually ends up in a really good place," declared Mark Zuckerberg back in the early days of Facebook, with an impressive combination of naiveté and disingenuousness. At worst, and dismayingly, this sees revolutionary moments recast as cultural shifts generated by disruptive thought leaders: history understood as the march of great entrepreneurial CEOs. This kind of thinking sees the future as defined by universal progress—rather than by a messy, contradictory struggle between different interests and forces—and never driven by the aspirations of those from below. It reduces the value of human agency to entrepreneurialism and empty consumerism.

History has a role in telling us about the present but not if we use a frame that valorizes those who currently hold positions of power. We need to reclaim the present as a cause of a different future, using history as our guide.

By stitching historical ideas and moments together and applying them to contemporary problems, it is possible to create a usable past, an agenda for an alternative digital future. In times gone by, early adopters, tinkerers and utopians may have wished for—even expected—a brighter and bolder future than where we find ourselves today, and I am keen to reclaim this possibility. This book will attempt to build bridges between technologists, activists, makers, and critical thinkers, to give shape to the "us" in the question "What is important to us?"

The histories in this book are stories of action, of revolutionary thinking but also revolutionary power in practice. They are also cautionary tales and stories of defeat, from which hope can spring eternal. "Knowing that others have desired the things we desire and have encountered the same obstacles," Brooks argued, "would not the creative forces of this country lose a little of the hectic individualism that keeps them from uniting against their common enemies?" Such an aspiration might similarly be extended toward readers of this book. The point is to use history as a guide for organizing and pursuing digital democracy collectively. On this foundation, we can start to build alternative visions of politics, law and technology.

The phrase "digital revolution" captures something of the transformative nature of the time we find ourselves in, but rhetorically also conceals the commonalities we share with the past. For this reason, it warrants a little explanation. Technology is revolutionizing how we organize production, reproduction and consumption. These changes also contain revolutionary political potential—though much of this remains unrealized—or struggles to find a form under capitalism. I therefore use the term as an accurate reflection of the changes brought about through the adoption of digital technology, but also with some skepticism regarding how fully the possibilities unleashed by this development have been explored.

We live in an age steeped in pessimism, in which phenomena like climate change threaten the lives of billions, inequality grows unchecked, and right-wing populism peddles fear and bigotry. The appetite for radical social transformation to address these trends is often lacking, but with notable exceptions. Left-wing ideas are still popular, and alternatives to capitalism are beginning to look feasible, promising and necessary. Radical proposals for universal government programs and redistribution of wealth have proven attractive to many in major social democracies. Developments in digital technology afford us some glimpses of how this might come about and how human ingenuity and cooperation have the potential to overcome profound challenges.

Marx claimed that revolutions are the locomotives of history. Revolutions transform how we live and work, junking ossified practices in favor of brighter futures. They generate an energy and change that drive us forward collectively, in a world where wealth and privilege might otherwise prefer slothful stasis. Yet, in our current age of rapid technological transformation, capitalism appears to be a constant, prioritizing selfishness over the immense human cost of greed, and squandering the potential of the digital age.

Walter Benjamin offered a reversal of Marx's proposition: revolutions might be the act by which humanity on the train applies the emergency brake. We need social movements that collaborate—in workplaces, schools, community spaces and the streets—to demand that the development of technology be brought under more democratic forms of power rather than corporations or the state. As the planet slides further toward a potential future of catastrophic climate change, and as society glorifies billionaires while billions languish in poverty, digital technology could be a tool for arresting capitalism's death drive and radically transforming the prospects of humanity. But this requires that we politically organize to demand something different.

If we are to explore the possibilities of digital technology, we need greater engagement between historians and futurists, technologists and theorists, activists and creatives. Synthesizing thinking across these fields gives us the best chance of a future that is fair. This is an ambitious project, especially at a time when the powers of capital and state are ranged against it. But as Vincent van Gogh reminded himself: "What would life be if we hadn't courage to attempt anything?"

2

An Internet Built around Consumption Is a Bad Place to Live

Cityscapes, as Imagined by Sigmund Freud and Jane Jacobs

Sigmund Freud has been described as "a frustrated archaeologist." The idea of cities and their ruins, and the layers of history they contain, was a metaphor that he returned to repeatedly in his writing. (On a personal level, he also collected thousands of antiquities, perhaps as a kind of performative psychoanalysis.) In the opening pages of one of his later works, *Civilization and Its Discontents*, Freud discusses the history of a city repeatedly built and rebuilt over time, using it to represent the human mind.

Rome, for example, became the global metropolis it now is over the course of thousands of years. People had lived in the area for millennia, and the city itself began as a collection of pastoral settlements on the Palatine Hill. As villages developed on various surrounding slopes, they formed a federation of sorts before a monarchy took control of the area around 800 BCE. This eventually transformed into a republic, which grew in power and influence and ultimately extended its empire across the Mediterranean.

The modern city of Rome looks very different from how it began. But any visitor to Rome today can see evidence of its ancient history everywhere. There are the old buildings that have lasted for centuries, as modern constructions rose around them. Beneath the buildings and infrastructure lie layers of rubble and detritus from generations past. Most spectacularly, the Forum displays its ruins for all to see, a rich seam of history through the center of the city. It gives solid evidence of the long and powerful reign of the Roman Empire, even if some imagination is required to conjure a vision of its glory days.

A city like Rome has inevitably evolved around certain definitive forces of nature, such as waterways and cliffs. It is possible to transform these topographical influences with the help of diggers and dynamite. Sometimes this kind of modification of the natural world is necessary for the residents, with road tunnels hacked through mountains and sewerage pipes buried underground, but in most respects humanity concedes to the landscape.

All this, Freud surmised, is not unlike the human mind. Historical experiences build up in our heads over time, with a persistent hold over the present. Some forces that create the topography of the psyche reside in the unconscious, like natural features of a landscape, resilient and sometimes difficult to control. Other influences are a result of the social context we exist in and respond to, like buildings dilapidated and rebuilt, determining the urban structure and layout. A comprehensive picture of the human mind requires proper analysis of all these elements and how they fit together.

Freud was quick to point out the limits of such a metaphor, and on other occasions he left its significance unstated or its meaning incomplete. A more precise analogy with the human mind, Freud insisted, would involve a visitor to Rome seeing it all at once—its past and its present, with buildings of multiple temporalities visible together. Real estate in cities is almost always in short supply, whereas the human mind knows no such material limitations, so the comparison can only take us so far. But there is something in the social complexity and the historical physicality of a city that gives the metaphor an enduring appeal. Freud's ideas, including his proposition that our minds work both unconsciously and consciously, and that they are a product of their history as well as their present, align neatly with the various influences that dictate how a city is built.

Many Freudian concepts are taken for granted today. But at the time he was writing, in the early twentieth century, they were a revolutionary approach to understanding the psyche. Questions about the intersection of agency, influence, pathology and plasticity within the mind were opened up by Freud in unprecedented ways. They are questions that help us consider how digital technology has affected our thinking and behavior. As our world becomes a place where we live and work with devices constantly at our disposal, it is worth thinking critically about how the evolution, design and regulation of digital technology bear upon our ability to build a rational psyche that is fulfilled, joyful and socially functional. Cities are planned in certain ways, to protect heritage, build sustainability and preserve amenity, so we can experience spaces differently according to our needs. They can also be designed for the purposes of social control, to limit particular interactions. Who exercises the power of design is a significant question. Given the enormous influence of digital technology on how our minds work, we need to find ways to make sure that it functions consistently with what is necessary to maintain a healthy and effective mind—the equivalent of a well-planned city.

Digital technology has put us in a bind. We have good cause to be suspicious of how much our personal engagements with it are influencing our behavior, but we are short of meaningful ways to challenge these effects. Or, as Walter Kirn put it, "if you're not paranoid, you're crazy." Companies are using digital technology to collect data about us on a larger scale than ever before, in ways that are far more revealing about our inner lives. It is like occupying a city where the topography changes daily. Fewer parks and gardens, more shopping malls; less mountainous horizons, more flashing billboards. Digital technology is creating a history of our sense of self that is tightly bound up with the market: the space of our mind is increasingly being defined by our consumption, for the purposes of further consumption. As digital technology develops at a dizzying pace, integrating itself further into our personal spaces, political communities and workplaces, each new advance is used to learn more about us at more intimate levels and map the contours of our psyche more intricately, often without our knowledge.

Framing is important here. Companies collect data, rather than—as is often claimed—we give it away willingly. Both constructions of the process are technically true, in the sense that the collection of our data is impossible without our formal consent, but that provides only a very limited picture of the phenomenon. Consent is in no way meaningful when online spaces are designed around the expectation that it will be given and rarely offer users an active choice in how their data will be used or managed. It is as though obtaining consent were a mere formality, secondary to another purpose.

Corporate surveillance is creating a robust social graph of our existence. To see this as a problem of personal, individual responsibility would be to miss the broader, systemic context. For this reason, framing these discussions around the idea of privacy—understood as an inherently personal right—falls short. The idea of privacy is too often a blunt substitute for notions such as agency, spiritual nourishment, and freedom from control. Privacy, as it is commonly understood and deployed, cannot help us understand how to avoid manipulation of both our unconscious and our conscience (or super-ego as Freud would put it); it is an inadequate description of a desire for space to build a functional ego that can rationally pursue a personal direction while simultaneously navigating the various compromises of living in a society.

The Internet is not just pipes and switches, and the web is more than hypertext; cyberspace is a place in which we engage with the world, and the forces that shape this place have influence on us. The structures surrounding our personal experience online are the focus of much commercial interest and investment, so much so that any meaningful struggle to create space for our inner self, for a literal form of self-determination, has to begin with an analysis outside of the self. No one would expect to find a moment of quiet reflection in the middle of Times Square or Shibuya Crossing. But everyone would be concerned if the spaces of beauty that might be perfect for this purpose—public gardens, monuments, art galleries—were being bought up and demolished to make more space for shopping meccas and strip malls. How we engage with the world on an individual level is deeply connected to the context we find ourselves in and the social forces it represents.

There is a growing awareness of the scandalously invasive way in which we are surveilled by data miners and marketers to predict and control our behavior for the purposes of making money. Cambridge Analytica is a good example. The company harvested data from Facebook via a personality quiz taken by tens of millions of people, and their friends, by stealth. Users, civil society groups and lawmakers were outraged that Facebook could treat our information so carelessly and that this information went on to become fodder for a powerful, profitable industry.

As we learn more about the ways in which we are identified, categorized and manipulated, technology races ahead, becoming ever more sophisticated and elusive. But while some of the invasive uses of digital technology can seem overwhelming, the battle for control of our destinies has not yet been lost to the age of the data boom—there is time to take charge of these processes before they take further charge of us. By learning and ultimately understanding, we can find ways to create space for our authentic selves in the digital age.

The idea that your phone is listening to you, or that your computer is watching you, is becoming accepted as a reality of life. To many of us, our personal data might seem deathly boring or perhaps individually embarrassing, but we increasingly recognize its value to unseen, shadowy forces such as data miners and marketers. Their activities do, after all, drive the economic powerhouse of Silicon Valley and explain how companies can be worth billions of dollars while offering us services supposedly for free. In the past, data collection and consumer analysis might have happened through subscription services or loyalty card schemes, something that was opt-in. But now that we spend large amounts of time online, whether at a computer or on a smartphone, the opportunities to collect this information have exploded. The pursuit of big data has created an army of cyber vampire squids, relentlessly jamming their blood funnels into anything that smells like it could be monetized.

Perhaps most iconically, this phenomenon caught the attention of the public when Target Corporation predicted a teen was pregnant before her own parents knew. Andrew Pole, a statistician for the company since 2002, was running tests on the data the company had on consumers to figure out how to leverage this for more effective marketing. Target had long collected customers' data and would send them personalized coupon booklets, based on past purchases, to draw them back to the store. "We do that for grocery products all the time," a Target executive nonchalantly told the *New York Times* for a story in 2012. As a statistician, Pole's objective was even more precise: find the right moment to market to consumers and change their shopping habits to win their loyalty to the store. His job was "to identify those unique moments in consumers' lives when their shopping habits become particularly flexible and the right advertisement or coupon would cause them to begin spending in new ways." Far and away, the most unique moment from this perspective is the birth of a baby. Babies represent a moment of flux: when old shopping habits crumble under the weight of caring for a new human, marketers have a precious chance to cultivate new behaviors.

Pole's tests produced a number of ways to predict when a baby was due, so that marketing material could target that customer before delivery and ahead of the competition. (Birth records are usually public; if Target waited for this data to initiate its marketing campaign, it would be lost among all the other companies doing the same.) Pole's model picked up the kinds of shopping habits that marked the beginning of a woman's third trimester. Hence the arrival of a whole bunch of ads for maternity clothes, lotion and diapers addressed to a high-school student, to the outrage of her father, who laid into the local store manager for inappropriately encouraging his daughter to fall pregnant. The father later apologized. It turned out the company had known something before he did.

These methodologies for predicting and shaping our behavior have grown more sophisticated over the first two decades of the twenty-first century. Collection and analysis of big data about people is a well-established industry. It includes the companies collecting data (miners), those trading it (brokers) and those using it to generate advertising messages (marketers), often with overlap between all three. It can be hard to obtain reliable estimates of the size of the industry, given its complexity, but one study says that by 2012 it was worth around $156 billion in the United States and accounted for 675,000 jobs. It has undoubtedly grown since then. Like slum landlords who rent out dilapidated apartments, or greedy hotshot developers who take advantage of legal loopholes to build luxury condos, companies that trade in personal data represent the sleazy side of how digital technology is impacting the real estate of our minds. This industry uses the faux luxuries of choice and convenience to entice us to part with our data, but often what they are really selling is overpriced and dodgy.

Via desktop computers or laptops, companies can collect information about us using a variety of methods, from what you click on to how long your mouse lingers. This involves the use of cookies, stored in the browser so companies can later track where you go by pinging messages back to the company servers that left them there. Such information can then be matched to other data sets, including financial information, purchasing histories and assumptions about health conditions.

Extracting personal data from smartphones is a more complex task than from personal computers. Unlike on a computer—where we often spend our time in a single browser that quickly accumulates cookies that can be analyzed—data on smartphones is often siloed into various apps, without much engagement between them. But this data is also much more valuable: we carry our phones with us on our person almost all the time, and each device has its own unique identification number. Retailers rely on beacons to pinpoint your location and provide what analysts describe as "a more customized approach to in-store shopping." If they can get their hands on your IP address, often via the operating system, marketers can know when you walk past a store. Lots of popular free apps also have standard terms that involve accessing significant amounts of personal data, including things like locational information, which is another way to know a user's whereabouts if the operating system settings are turned off. These apps (flashlight apps are perhaps the most notorious) act like digital Trojan horses, with users downloading them without realizing they are a back door for data miners. The race to improve the capacity of smartphone technology to provide a richer picture of user behavior represents an important frontier for the data boom, and it is likely to be a continuing focus into the future.

This kind of tracking is also happening between devices connected to the Internet—from television set-top boxes to smart fridges. One method for cross-device identification involves the use of ultrasonic frequencies inaudible to humans. The sounds play in television or browser ads and can be picked up by smartphones. This can track what ads a person sees and, by linking devices, whether they respond to the messaging by searching for or buying a product. The potential of smart televisions to use data collected from your different devices, combined with independent data sets, is transforming the idea of personalized ads, something the advertising industry is highly enthusiastic about.

These advances make it very difficult to meaningfully know, let alone limit, what information is known about you by others. The technological infrastructure around our online behavior and our physical presence is constantly observing what we do. As the Center for Digital Democracy describes it, this cross-referencing "has effectively erased any privacy safeguards we may have enjoyed previously when we switched between devices." Switching browser programs used to be another effective way of preventing data miners from collecting a complete picture of your activities on a computer, but this is changing. Browsers perform a range of tasks as we use them, loading graphics and using plugins, for example, and computers will have certain settings, such as a time zone, all of which are specific to individuals. Taken together, these can allow a user to be identified with over 99 percent accuracy.

The industry is also adopting various forms of biometric profiling, including using keystroke patterns. How we type is marked by minute differences, which can create a biometric profile of individuals and even be matched to emotional states. Thus you might be identified even when using an anonymized browser, such as The Onion Router (Tor), or when using a different computer. Facial recognition software is already prevalent in the retail industry, and increasingly it is being matched to other data sets and used with beacons to push advertising onto smartphones. Over time, marketing that connects our different behaviors together—on the street, in the home, at work—will be the norm.

All this together means that the data mining industry is enormous and increasingly difficult to evade. While it is easy to see how Facebook and Google obtain enormous amounts of our data, there are dozens of more shadowy companies that do the same on an even larger scale. These companies are not household names, yet they hold countless intimate details about us on their digital ledgers. We find ourselves walking through a city of private eyes and pickpockets, constantly watched by closed-circuit cameras that fade into the backdrop of urban life. This intensive examination and analysis of our behavior by private companies is what Shoshana Zuboff has called "surveillance capitalism"—companies snooping on us for the purposes of selling things. It sits on the very edge of technological development, sucking resources into its projects, and often involves obscure companies as well as mainstream platforms. Together the industry holds billions of pieces of information on billions of people. As the advocacy organization StopDataMining put it: "If iron ore was the raw material that enriched the steel baron Andrew Carnegie in the Industrial Age, personal data is what fuels the barons of the Internet age."

One of the problems that arises in these discussions is metonymy: privacy is a single word that describes all manner of dastardly approaches to information management. It is a term often used by companies in transactional, technocratic ways, ignoring the conceptual and material implications of their approach. Secrecy, security and anonymity all describe slightly different components of privacy but regularly end up lumped together. Secrecy covers the confidentiality of a communication—that is, secrecy exists when the substance of a message is known only to the sender and the intended recipient. Security is about the integrity of communication channels and certain spaces (be they physical or cyber), to ensure they are free from invasion by uninvited parties. Perhaps most importantly for our present purposes, privacy is often used interchangeably with anonymity—that is, where information collected is separate from the name of the person it is collected from. Many companies that talk about privacy only offer in reality one, maybe two, of these guarantees. Such a promise is a fudge. Or as the Center for Digital Democracy puts it: "This is merely a 'don't-look-too-closely' claim designed to head off the scrutiny their practices require."

There are two reasons for being skeptical of the promise of "anonymous" or depersonalized data. First, on a practical level, anonymity is brittle. It is easy to identify someone using only a few data points. Way back in 2000, professor Latanya Sweeney found that 87 percent of Americans could be uniquely identified using only their ZIP code, gender, and date of birth. So, for what it's worth, the more information companies collect and hold, even if it is held nominally separately from our name, the easier it gets for someone to reverse-engineer this to link the data to us. To protect ourselves from harm, we are dependent on companies protecting our data, even if it is de-identified or innocuous, at a time when leaks and hacks are commonplace.

Second, on a more abstract level, the protection of privacy offered by anonymity alone is minimal. Companies create identities for us based on this data, without any accountability or capacity for us to change them. Our *abstract identities*, if we understand that to mean our social, political and economic preferences as determined by the data collected about us, are generated and then repeatedly refined and used to determine advertising for us. Data points lead to assumptions about relevant marketing, which lead to further data points that make up your abstract identity. It creates a form of path dependency: once the vast and obscure apparatus of data-driven advertising takes a particular course, the choices it makes about you are constrained by its previous choices. These abstract identities follow us around online, even if they are not attached to our name, like zombies. They are beyond our control. In this light, the absence of a name attached to that identity offers scant protection from anything meaningful. Data collection and curation, even when done in ways that protect our anonymity, limit our freedom individually and collectively.

This process of *abstract identification* curtails autonomy by creating a summary of your personality—thoughts, needs, desires, and especially vulnerabilities—extracted from data generated online. Using the highly selective body of information, this process creates a history of your sense of self that serves to influence you. You are stripped of your agency; lacking the capacity to control what is known about you and by whom, your ability to make decisions for yourself is impaired.

Autonomy, then, is the other essential aspect of privacy that is rarely given its full meaning in mainstream discussion of the topic. Secrecy, security and anonymity are all important, but autonomy is too often ignored, reducing privacy to a transactional concept, depoliticizing it and confining it to the atomized individual. Allowing others to write a history of our sense of self forecloses the possible futures available to us. "Privacy is the right to a self," declared the whistleblower Edward Snowden. "Privacy is what gives you the ability to share with the world who you are on your own terms." It is the bridge between the individual and the social, between our selves and our context. To give the idea of privacy the richness it needs in order to be meaningful requires that we understand it collectively, as a function of power.

Many data miners and marketers are not concerned with who you are in the real world, but they are highly interested in your abstract identity (your suburb or city, for example, or your car model and make). They are not bothered about linking this to an actual name or physical presence. For the most part, they are only interested in you as a consumer, someone who buys and who can be convinced to buy; a data point that fits into collective trends or cohorts of people who behave similarly, someone who can be predicted to behave in certain ways in response to particular stimuli.

Freud's thinking gains new relevance in this context. By understanding that our mind is made up of both a conscious and an unconscious, we can start to appreciate how our existence in the digital age is not just a matter of choice, nor is it simply driven by our own free will. Freud explored the idea that mental processes are driven by both the pleasure principle and the death drive. Our minds, he argued, are motivated by the "production of pleasure," an observation that is easy to identify with. But we are not all simply hedonists, and to some degree we must temper our desire for pleasure with the limitations of living in a society, what Freud called the reality principle. And, he said, we also possess a self-destructive tendency, or death drive, manifested perhaps most prominently in survivors of trauma or pain who repeat thoughts and actions associated with those experiences. This comes from a desire to overcome that past event: "to work over in the mind some overpowering experience so as to make oneself master of it." These kinds of influences on our mental processes can be subject to manipulation or intervention, be it by the analyst or by others with less therapeutic motives.

In the digital age, these features of our psychology are subject to manipulation in all sorts of ways now that our social interactions and material consumption increasingly occur online. As marketing budgets grow and companies spend more on mapping the content of our abstract identities, it starts to look like a hopelessly mismatched battle of wits. It is not that we are all dupes. When marketers know both what kind of pleasure we desire and what kind of self-destructive habits we practice, it gives them an enormous amount of power in a context in which many of these behaviors find expression online.

Freudian ideas, and the broader body of thought around psychoanalysis, provide insight, even hope. They encourage us to appreciate the power of the individual to come to know what is unknown, to identify manipulation even when it might be well concealed. Psychoanalysts argue that we have the capacity to make conscious what we have hidden or repressed in the unconscious—trauma, forbidden desire or other experiences. And the unconscious exists as "neither individual nor collective," writes the philosopher Mladen Dolar, but rather "precisely between the two, in the very establishment of the ties between an individual (becoming a subject) and a group to which s/he would belong." In other words, there is a dialectic process at play between the social forces that shape us and our own personality. While the data mining industry might seek to make use of this for its own commercial ends, resistance is not impossible.

A city is a conscious attempt to collectively dominate nature—to build onto the natural world so as to protect citizens against the elements. But there is also an impulse to maintain something of the natural green spaces in our urban environments, in an effort to keep cities sustainable and perhaps remind us of the vast and intricate network of life that exists in the land, water and sky all around. We will always be required to mold our temperaments to accommodate the experience of living in society, but our minds need room to breathe and space to explore possibilities of independence and collaboration, free from corporate agendas. Finding the right balance between the constructed and natural environments is a challenging task, and this holds true for our psyches also.

The most valuable consumer platforms have both the capacity to collect highly valuable personal data and the opportunity to use it to market to users at the most lucrative moments of their daily lives. These are the places in which the invisible hand of what I call *technology capitalism* is at work—between data miners and advertisers, with data on users as the commodity being traded.

For our present purposes, I will define technology capitalism as the leading edge of the technology industry, a system led by a class of people who are focused on orienting digital technology toward market-based systems of profit. My aim is to use the term to demarcate the active parts of this modern industry, rather than use it as a generalized description of capitalism as transformed by technology.

These platforms are the places in the digital age where personal data is centralized and then used to segregate us into different audiences of consumers. Amazon has a record of all you have purchased and everything you view and uses it to generate specific ads and differential pricing based on where you live and other personal information. Google has a log of all your search histories, your emails, your YouTube views, your Nest data, and your locational information using maps, which it uses to generate personalized advertising. Every Facebook like and share button you see is tracking you, even if you are logged off Facebook, regardless of whether you click on the button. (These buttons are actually small pieces of code, which instruct the browser to contact Facebook's servers when you land on the site.) The same is true for websites with embedded YouTube videos, which feed data back to Google. This all feeds into a database of your habits and interests. Sometimes these companies also purchase data to bulk out their own pool. Together they have highly sophisticated and well-used platforms that can be relied on for creating a picture of your abstract identity.

This system of observational intelligence, scattered across the web, is then used to curate our singular sense of self, that is, our real-world understanding of our own personalities. "The advantage that Facebook and Google have over the regular data on-boarders is twofold," writes Antonio García Martínez, a former product manager in the Facebook ads team. "They have much more of your personal data, and they see you online all the time." That is, they have the capacity to collect rich and diversified data and to then use it to deliver compelling marketing messages. "Facebook, Google, and others have achieved the holy grail of all marketers," he continues. "A high-fidelity, persistent, and immutable pseudonym for every consumer online." The more a marketer knows about you, the better the ad; and the more time you spend on a particular platform, the more valuable the digital real estate.

The precision is impressive. A ProPublica investigation found that Facebook offers advertisers more than 1,300 categories of users to allow them to carefully direct their messages: "everything from people whose property size is less than .26 acres to households with exactly seven credit cards." Your own history of your sense of self, when documented by these platforms, works to determine your future. This is how technology capitalism dominates our personal experience of digital life. We might think we are in a public square or community-owned garden, but in reality we are not living in the same exact city as anyone else.

For these reasons, no two people's experience of the web will be the same; websites are put together based on our abstract identities. Professor Joseph Turow compares this to selling by peddlers in centuries past, versus the experience of the shopping mall. Shopping malls were a space where everyone had access to the same goods, for the same price. Peddlers—the old way of buying and selling—traveled door to door, sizing you up according to the looks of your home, remembering what you bought from them last time. "Peddlers evaluated you based on their relationship with you, they changed their prices based on what they thought you could afford," says Turow. "There was all this negotiation back and forth." In the wake of the digital revolution we are returning to the peddling mode, where products are presented to you based on information about your preferences, status and vulnerabilities, and prices are increasingly set for each specific consumer. As Michael Fertik, the founder of Reputation.com, bluntly put it: "The rich see a different Internet than the poor." With the move away from the public space of a mall or market square to individually tailored transactions, Turow argues we are experiencing "a major transformation of what it means to buy and sell in the public sphere." It challenges the very idea that the Internet is a public space.

The justification for this process of abstract identification and segregation is often framed in the nebulous discourse of consumer choice. Jeff Bezos, CEO of Amazon, reportedly reserves a seat at the conference room table known as "the empty chair." At meetings, Bezos reminds attendees that this represents "the most important person in the room," namely the customer. Data collected by social media or otherwise is often presented as a way of giving consumers access to relevant advertising and offers that may be of interest to them. Indeed, one of the largest data miners, Acxiom, has encouraged people to look up the data it holds about them to correct any mistakes. In other words, Acxiom was asking people to more actively engage in the process of abstract identification—a highly valuable, low-cost exercise for them.

In this way, data mining is framed as a supposedly evenhanded power relationship in which consumers are invited to dictate to large corporations how they want to live their lives online. We are told that the process of abstract identification is something that benefits us all, individually and collectively, that it can even be empowering. This is an effective but misleading portrayal of the dynamics at work. It is not unlike a city authority permitting the razing of historic districts or lively green spaces to make way for the construction of parking lots and luxury stores. While we all get to walk past and peer into these expensive buildings—and theoretically we all have the freedom of choice to make use of them—they occupy spaces that previously held relevance for all residents and now benefit only a select few. They frame our expectations around the meaning of success and pleasure. Such projects are a great outcome for property developers and their wealthy clients, less so for the rest of us.

Surveillance capitalism yokes us to a certain history of our sense of self, which is constantly updated and channeled into a vision of our own future. Collectively we rehearse the experience of being a consumer in all our communities every day, and we practice letting companies collect our personal information, so that it feels normal. We learn to relinquish the idea that there should be spaces free from this dynamic. We grow accustomed to understanding our psychological real estate as a resource for fueling capitalism, rather than a social or personal space. It is the twenty-first-century equivalent of Margaret Thatcher's favored slogan TINA—There Is No Alternative—that deflects us away from imagining other ways of structuring our digital lives. In his work as a psychoanalyst, Freud saw firsthand how self-knowledge—the expression and analysis of our mind in therapeutic ways—could help us navigate the challenges of living in society. The current structure of online life not only militates against that objective, it hands over deeply personal information to companies that are not interested in helping us to overcome our problems or develop a functional mind. The companies do not need us as consumers; this is a world in which we are socialized into needing them.

Rather than building connections, let alone a genuinely public space in which people can communicate collectively or buy and sell in the marketplace on equal terms, the companies that profit from the data boom are breaking these spaces down. Data is centralized, with the effect of creating specific and segmented populations based on demographics. "The goal is to distribute, not concentrate, the population," argues professor Bernard E. Harcourt, "to avoid amassing consumers at any one spectacle—so that they spend much more at all the various mini-theatres of consumption." As Turow sees it, the kind of marketing that we are subject to alerts us to our social position: "If you consistently get ads for low-priced cars, regional vacations, fast-food restaurants, and other products that reflect a lower-class status, your sense of the world's opportunities may be narrower than that of someone who is feted with ads for national or international trips and luxury products." He calls these "reputation silos," which have the effect of entrenching the distance people feel between one another. In other words, shared public space begins to vanish; increasingly there is no single collective experience online. The age of the all-powerful consumer is actually one in which we are offered narrower and narrower choices, assigned an identity that reflects, and continuously reproduces, the way in which technology capitalism is shaping our experience of the web. It is a version of what sociologists call "symbolic interactionism," which describes how individuals and societies are constantly producing and reproducing identities and norms through social, or in this case digital, interaction.

If we think about the Internet as a place rather than a service, this process of abstract identification is not empowering—it causes fragmentation and distance between people. It creates a world where the population is subject to different framing effects, making for increasingly insurmountable political and social divisions, worlds that stand apart from each other despite nominally existing in a communal space. Gated communities become sealed off from socially isolated public housing projects. The effect is to degrade our sense of belonging, to emphasize the differences in our abstract identities rather than our commonalities. "By emphasizing the individual to an extreme," Turow writes, "the new niche-making forces are encouraging values that diminish the sense of belonging that is necessary to a healthy civic life."

The iconic activist and urban studies writer Jane Jacobs described the importance of understanding "what makes a city center magnetic, what can inject the gaiety, the wonder, the cheerful hurly-burly that make people want to come into the city and to linger there." Jacobs came to prominence in the 1950s and '60s, defending Washington Square Park in New York City from plans to demolish it to make way for an expressway. Her most famous and influential book, *The Death and Life of Great American Cities* (1961) argued that one of the key principles underlying successful cities is the presence of an "intricate and close-grained diversity of uses that give each other constant mutual support." Jacobs rejected the modernist, rationalist design that dominated orthodox urban studies at the time, in which the natural direction of urban renewal was toward planned cities and high-rise buildings, rarely integrated into their surroundings. Jacobs embraced the value of landscapes that allowed people from different backgrounds, with different purposes, to all mix together, and advocated for building cities around this objective.

When it comes to the cities of our psychology, technology capitalism has adopted the mantra of the orthodox urban studies aficionados. The facilitated segregation and curation of non-intersecting worldviews shape a psychological landscape that is stripped of diversity. Jacobs would no doubt shudder at the prospect of such banality and isolation.

The fragmentation of our online public spaces and the way in which our entire sense of self has become highly porous to influence may have corrosive effects, but that does not mean it is an unpleasant process. Quite the contrary: it is designed to be enjoyable, structured to optimize, at times, a sense of fulfillment. Surveillance capitalism uses our desire for convenience and connection as bait to draw us into using its platforms. It then uses our consent to justify transferring responsibility for its invasive practices onto us. It is akin to organizing an entire city around cars—demolishing neighborhoods and parklands to facilitate the smooth flow of traffic. Parking lots and spaghetti bridges are convenient for drivers, insulated from weather and noise in their bubbles of steel and glass, but they also deprive us of the capacity to experience the charm and surprise of being a pedestrian.

A metaphor frequently used to understand surveillance capitalism comes from George Orwell's novel *1984*, where the all-seeing, all-knowing Big Brother watches our every move through hidden cameras, screens and recorders. Similarly, it is compared to Jeremy Bentham's nineteenth-century model of an "enlightened" prison, the panopticon, where the guard is able to see all of the prisoners all of the time from a central vantage point. Yet both these comparisons fail to capture some important features of our current predicament.

Big Brother and the panopticon worked on the basis that they taught people to police themselves. Surveillance capitalism is generally designed to go unnoticed. While many companies have a vested interest in mining our data and using it to influence us, this tends to work better the less we know about it. As Harcourt explains: "The watching works best not when it is internalized, but when it is absentmindedly forgotten." Target, for example, learned this lesson when the story of the pregnant teen and her unsuspecting father made headlines. Such marketing methods appear unnerving and intrusive, rather than impressive.

Mindful of this, plenty of websites operate as data miners in disguise. Dating websites are a good example. The sales pitch of OkCupid is that it takes a scientific approach to matching people, based on survey questions about values and preferences across a range of fields. But what the site does not explicitly tell you, or would rather you forget, is that any data you enter about yourself belongs to them. The company has admitted it has no idea of the relevance of the collected data to the objective of engineering better matches. They basically make it up as they go along by running experiments on their users. As a co-founder of the company put it, rather airily: "Guess what, everybody: if you use the Internet, you're the subject of hundreds of experiments at any given time, on every site. That's how websites work." The money is in the data that they collect and sell to third parties. To this end, OkCupid's parent company, Match Group, owns multiple dating sites—everything from Tinder to CatholicPeopleMeet—that cater to different races, religions, ages and political preferences. It is a form of abstract identification sold as an optimized chance at romantic success.

Marketing messages increasingly hide below the radar, barely even registering with many of us as advertising. In *Black Ops Advertising*, Mara Einstein talks about how Red Bull became a market leader in this area, with countless videos of daring stunts and entertaining feats that seem to bear little relation to the energy drink. Such videos are beautiful and exciting to watch. Only small clues mark them as advertisements, such as the placement of the logo. The subtlety of the marketing is what underpins its appeal. "Years of remote controls, DVRs, and now 'banner blindness' and ad blockers have taught advertisers that consumers are utterly adept at circumventing advertising," Einstein writes. "In response, they have turned to new and improved forms of clandestine marketing." The goal is to influence us to remember brands and imagine buying things without being aware of it. Ultimately, these companies want us to replicate this process by sharing the content and its covert branding, so our entire social network becomes a permanent marketplace. It is a way of commodifying and monetizing our social spaces.

This is Freud's pleasure principle writ large—a social experience that is centered around pursuing endless saccharine indulgences with the promise of avoiding reality. The psychological process that Freud identified has been repurposed to suit the aims of capitalism in online spaces. "The programme of becoming happy, which the pleasure principle imposes on us, cannot be fulfilled," wrote Freud, "yet we must not—indeed, we cannot—give up our efforts to bring it nearer to fulfillment." This can take different forms, such as the attainment of pleasure or the avoidance of displeasure. But no matter what we do, Freud noted, "by none of these paths can we attain all that we desire." The pleasure principle drives our behavior, and unless we find ways to manage its insatiability it can become dominant, even overwhelming. Surveillance capitalism has structured our online life so that attempts to limit the influence of the pleasure principle are made difficult and frustrating.

There are many troubling parallels with the electronic gambling machine industry. This industry is built on active and sophisticated attempts by technology designers to create a dynamic of addiction between person and device. Decades before the advent of data miners and marketers, technologists in the gambling industry were pioneering methods of amassing large amounts of personal data about users, cultivating consumer loyalty and finding pain points and precisely timed moments to deliver effective marketing messages. Cultural anthropologist Natasha Schüll has written about how engagement with gambling machines is "a trapping and ultimately annihilating encounter." Venue operators regularly use real-time monitoring of the play of individual consumers—turning gambling machines into surveillance devices—to orchestrate interventions that will keep people engaged. Gamblers become lost in what they call "the zone," with endless play generating the precise balance of stimulation and calm that locks them into subservience to the pleasure principle.

As is the case with data mining and black ops marketing, the industry behind gambling machine design is shadowy, almost unseen. Enormous effort goes into cultivating the kind of addiction that sees gamblers destroy themselves. Schüll documents "the painstaking efforts" by the industry "to organize the kinaesthetic and temporal elements of machine play." Such efforts underscore the fallacy of framing addiction in the language of individual responsibility and willpower. Rather, the cause of addiction is not discretely within the person or the technology but in the dynamic interaction between the two.

The same observation can be applied to surveillance capitalism: digital technology is designed carefully to foster continued engagement that serves the purposes of data miners and black ops marketers. Martínez writes about how the Facebook growth team, responsible for increasing user numbers, used the very same strategies employed by gambling machine designers. The growth team

> exploited every piece of psychological gimcrackery, every tool of visual legerdemain, to turn a pair of eyeballs into a Facebook user ID ... They calculated statistics like clickthrough and conversion rates out to three decimals, and maintained comprehensive databases of user data. Whether via Skinnerian or Pavlovian psychology, they'd figure out the optimal rate to send reminder emails about in-Facebook events (like mentions or new posts from friends) for optimal response.

It is not just social media platforms that use this strategy. Adam Greenfield has written about how we live in a networked condition, whereby the immense functionality of devices like smartphones encourage endless time spent swiping and scrolling. "We become reliant on access to the network to accomplish ordinary goals," he writes. In doing so, we put ourselves at the mercy of devices which constantly extract data and use it to manipulate us with tailored marketing.

For many of us, this is our daily experience of computing and the web. Living online is now such an overwhelming and undeniable experience that there is growing public discussion of how design cultivates addiction behaviors toward digital technology.

People like Tristan Harris from the Center for Humane Technology have critiqued design features that provide intermittent rewards, create endless newsfeeds, and manipulate our "fear of missing out" and desire for social approval. Harris has a tendency to frame this issue in individualistic terms, at times leaving the broader context unquestioned. His status as a former insider—he worked at Google—means he speaks with undeniable knowledge, but it does not necessarily equip him to provide the necessary critique of technology capitalism. Still, we should listen to such testimonies about the enormous resources being devoted to getting us online, to tune in and drop out. They are vital to understanding the dynamics of the digital spaces we have to navigate every day.

Professor Harcourt describes Facebook's user experience as "the digital equivalent of the perfect hallucinogen, 'soma,' from Aldous Huxley's *Brave New World*—a magical substance without side effects or hangovers, that is perfectly satisfying." Soma is perfectly satisfying to the point where life without it seems troubling and even miserable. Martínez puts it even more bluntly, reflecting on his decision to join the company: "Facebook was legalized crack, and at Internet scale." In this way, the analogy of Orwell's *1984* becomes unhelpful for understanding the phenomenon of abstract identification and data discrimination. As Neil Postman argued back in 1985, Las Vegas, with its slot machines and devotion to entertainment, was the best metaphor for our collective character and aspiration in the age of television—and, I would suggest, the age of digital technology. Rather than the tyranny of fear and hate that characterized Orwell's dystopia, ours is one ruled by the smiling face and its vaudeville entertainment. "Orwell's prophecies are of small relevance," Postman concluded, "but Huxley's are well under way toward being realized."

Freud appreciated that living in society is always a compromise between the desire to live a life of pleasure and the realities involved in living as part of a collective. Analysis could help us navigate around the trauma and pain that have befallen us, and not let our past determine our present or future. But we also have to survive in a social context and find ways to come to terms with how this inevitably tempers our behavior. This is not just an organizational burden but also an essential part of being human. "The individual only makes sense as a knot of social ties, a network of relations to the others," writes Dolar. Outside the social context we are meaningless, even if it also acts as a burden upon our desires.

The ways in which we build our spaces, both physically and digitally, will influence our capacity to manage our desires and navigate the compromises we must make. They should balance convenience with engagement, and cultivate connection, rather than atomize and segregate. These are exactly the motivations that drove Jane Jacobs to lead a campaign to save Washington Square Park in New York City from a plan by urban designers to build a freeway through it in the late 1950s. The proposed freeway, while certainly convenient for cars, would have destroyed a resource that was used by local people in a granular way, rarely appreciated by planners with grand, modernist visions. "It is very discouraging to do our best to make the city more habitable," wrote Jacobs to the mayor, "and then to learn that the city is thinking up schemes to make it uninhabitable." Ten years later, after countless meetings and protests involving thousands of residents, the plan for the expressway was scrapped. What had been dismissed by senior urban planners as a movement of "a bunch of mothers" had actually saved a part of the city that remains deeply important to its inhabitants.

There is much still to be won and lost in the battle for our online autonomy in the future. As the next generation of web technology improves the integration of all our digital activities, allowing machines to organize even more of our lives, others will continue to learn more about us than we even know ourselves. In this context, focusing on our power over this process as consumers is a mistake: the power being exercised over us is precisely based on our being socialized as consumers. This prepares us to accept a city where every park is paved over to build freeways and every sports field and roller-skating rink is demolished to build a shopping mall. Such a city would not be a functional, let alone enjoyable, place to live. But it would be a place where data traders and retailers made a lot of money.

To talk about this process as one conducted with our consent or understanding diminishes the immense effort that goes into facilitating our participation in such operations. It focuses the problem on the individual, rather than the system. It ignores the genuinely held concerns of many about the negative way in which surveillance capitalism affects their lives, and the widespread desire for something different. One of the first research scientists for Facebook, Jeffrey Hammerbacher, famously captured some of this disappointment: "The best minds of my generation are thinking about how to make people click ads," he said. "That sucks."

When Uber first launched in 2010, there was considerable fanfare about how ride sharing was a positive development. It would allow more people to work their own schedule; it would improve transport services to areas traditionally underserved by taxis; it would be environmentally friendly by discouraging car ownership. Uber was intended to complement public transport options, not replace them. In 2015, its then CEO Travis Kalanick announced the company's "simple" goal: "to take 1 million cars off the road in New York City and help eliminate our city's congestion problem for good." It was a classic moment in Silicon Valley optimism, epitomizing the promise of the digital revolution according to technology capitalism.

They were bold claims, and the upshot has been ambiguous at best. Rather than getting cars off the road, the rise of Uber appears to correlate with an increase in traffic congestion. Average travel speeds in central Manhattan have declined 15 percent from 2010. It is not entirely clear why this is happening, but experts think that Uber and other ride-sharing companies are part of the reason. The number of subway riders has dropped, despite population growth. Revenue for the city has also fallen: yellow taxi trips incur a 50-cent surcharge that funds improvements on subways and buses. In other words, the popularity of Uber has meant that public transit options have struggled to remain competitive. London fares no better, with more Uber drivers than black cabs now on the roads, reversing much of the environmental and practical gains made by the introduction of the congestion charge.

Uber may be convenient, more enjoyable than taking a packed bus, and cheap in relative terms. But the privatization of mass transit is also making our cities worse. Uber's long game appears to be to dispense with drivers altogether and run a fleet of driverless cars that offer even cheaper rides. It is easy to imagine our cities in constant gridlock, with public transit fallen into disrepair and more public spaces eaten up by widening roads. We will be chauffeured through the smoggy air, logged and tracked under the watchful eye of Uber HQ. From a city planning perspective, it is a world that profits Uber and not many others. This phenomenon is not unlike what is happening to our psychological landscape, as a result of the data mining industry: the cities of our mind are being clogged up by corporations with little interest in planning infrastructure and building skylines to suit our needs, but that prefer to prioritize their bottom line.

One of Jane Jacobs's key insights into what makes cities great was the importance of pedestrians. She felt that cities were best understood by walking around them, and she believed lively streets were critical for creating diverse and vital communities. From this perspective, we can start to see something of an alternative future. It could be to create a psychological topography that welcomes the flâneur, the iconic urban explorer and stroller of the streets who recurs in much literature and philosophy. The French writer Charles Baudelaire described this proverbial character as one whose genius is grounded in curiosity, who is gifted with the capacity to see. "To be away from home and yet to feel at home anywhere, to see the world, to be at the very center of the world, and yet to remain hidden from the world," Baudelaire explained, were some of the pleasures of the flâneur's spirit and independent nature. Later iterations, most notably from Walter Benjamin, highlighted the sensory overload created by consumer capitalism from the perspective of the flâneur, hollowing out his experience.

This perspective might hold promise for escape from the drudgery of a data-driven life. For the flâneur to occupy a city—to be enticed to explore it—implies a site of beauty, complexity and surprise. An element of the unexpected, the possibility of interesting encounters, rather than gated communities or soulless high-rises saturated by surveillance; a space cultivated around the needs of people, rather than profit. Can we aim to rebuild such a city psychologically, in resistance to the designs of the data mining industry? Jacobs observed that "every downtown can capitalize on its own peculiar combinations of past and present, climate and topography, or accidents of growth." Ultimately she was excited by the prospect of doing so:

> What a wonderful challenge there is! Rarely before has the citizen had such a chance to reshape the city, and to make it the kind of city that he likes and that others will too. If this means leaving room for the incongruous, or the vulgar or the strange, that is part of the challenge, not the problem. Designing a dream city is easy; rebuilding a living one takes imagination.

We need to protect space in our minds for the vulgar and the strange, for the unpredictable experiences of living free from the influence of commercialism. Like the flâneur or flâneuse, we should aim to cultivate curiosity and the capacity to see our fellow citizens through this liberated lens. The flâneur "participates fully through observation," as one writer put it. If we embrace that mode of being, we can begin to see how we ourselves are being observed by the structures of surveillance capitalism—and stroll with determination toward the objective of dismantling them.

3

Digital Surveillance Cannot Make Us Safe

Policing Bodies and Time on London's Docks

London in the 1790s was a filthy and bustling metropolis of empire. Maritime traffic up and down the Thames was growing at an astonishing pace, with riches plundered from the colonies winding their way toward markets through this central artery of the city. In a bid to speed up the process of docking and unloading, the West Indian merchants pooled their funds to build a wet dock on the Isle of Dogs in the east of the city, freeing them from dependence on the tides. But this logistical improvement exposed another feature of industrialization. The historian Peter Linebaugh has written about how workers in manufacturing and production houses often supplemented their wages (if they were lucky enough to be paid any) by pilfering stock from their employers. It was customary to do this, especially at a time when workers were testing the limits of the idea of wage labor—a distinct change from their previous ways of life as artisans or peasants. Linebaugh documents social and even judicial recognition of this practice, whether it was warehouse laborers drinking the rum maturing in the storehouses, a printer retaining a copy of every book he assembled, or shipbuilders taking leftover timber for their own use. It is not unlike how a contemporary retail worker might be given a discount on clothing to wear during her shift, or a café waiter given excess pastries at the end of the day. Or perhaps like start-up employees owning a share in the company as part of their salary. In the London of those days, this culture represented "a kind of 'collective bargaining' over the materials of production," Linebaugh argues.

Among the merchants, however, it was increasingly thought of as theft. The construction of the wet dock put the river workers under increased scrutiny—the process of unloading sugar and other commodities could now be carefully tracked and measured between when it arrived on the ship and when it was eventually sold. The merchants wanted the tradition of workers helping themselves to be stamped out: they were to work as wage laborers paid for their time, and nothing more. But it was a slog getting the slow and unwieldy arm of the law to enforce this.

Until that point, London had no police force. While constables and night watchmen existed in various areas, these bodies were diffuse, unruly, unprofessional, underpaid (if paid at all) and often corrupt. There was no professional body of officers under state control, authorized to use force.

In a direct attempt to protect their property and corral the dockers to conform to the relatively novel strictures of wage labor, John Harriott, a magistrate, farmer and businessman, came up with a plan for a professional police force. He was not the first to imagine such a thing, but his proposal was nonetheless unprecedented. He teamed up with Patrick Colquhoun, a merchant and statistician of repute, and Jeremy Bentham, the famed utilitarian philosopher. Colquhoun and Bentham brought their prestige and political judgment to Harriott's bold ideas, and together these men sought to design organs of civil government that could address the challenges of the industrial age. They wanted to figure out how collaboration among the propertied classes would best allow them to maximize the opportunities presented by the novel and developing system of capitalism.

Specifically, this resulted in a proposal to the merchants' committee to fund an experiment, which was duly accepted. The Marine Police Office opened in 1798 as a kind of pilot program. Officers were paid and uniformed, and their job was to watch over the wet docks. They supervised the workers and kept an eye on the ships and their cargo. They enforced working hours and were even responsible for paying out wages. When they encountered misbehavior, they did what was necessary to bring the errant workers before a magistrate.

It was a raging success, significantly reducing the merchants' losses for a small price. Harriott congratulated himself for "bringing into reasonable order some thousands of men, who had long considered plunder as a privilege." Working their magic in a way that would impress even the toughest political lobbyist, Harriott, Colquhoun and Bentham managed to convince Westminster to support the project, and the Marine Police Office came under state authority two years later. Not only did the merchants have a police force, it was now funded by the public purse.

Colquhoun wrote about his experiment on the River Thames extensively, keen to export his model around the world. His vision of policing was deeply bound to economics, and he saw the urgency of promoting this model as the industrial revolution transformed material relations. "Police in this country may be considered as a new Science ... in the PREVENTION and DETECTION OF CRIMES," he declared a few years later in his treatise on the topic. "Under present circumstances of insecurity, with respect to property, and even life itself, this is a subject that cannot fail to force itself upon the attention of all."

The eighteenth century was a time in which the contours of public life and governmental authority were being defined in ways that would last for centuries. How best, then, to inculcate a culture of compliance with this new way of doing things? "It is the dread of the existing power of immediate detection, and the certainty of punishment as the consequence of this detection," wrote Colquhoun, "that restrains men of loose morals from the commission of offences." Surveillance, plus the spectacle of force, offered an economical and effective way of maintaining order. It is not walls, locks or bars that prevent crime, Colquhoun argued; instead, "restraints are only to be effected by the strong and overawing hand of power." He was describing a tactic that would be deployed by the powerful in our own era: surveillance as a method of social control.

Colquhoun determined that the conversion of river workers into a disciplined, regulated class of wage laborers could only be achieved through moral transformation. This transformation involved the criminalization of idleness and a revision of the sense of injustice that made workers feel entitled to a decent share of the fruits of their labor. It involved the pursuit of order, and faith in the state to protect people from the threat of disorder. Without an organized defense of private property, the nascent capitalist system would inevitably fall into disarray. This requires a certain form of subordination. "Civil government," observed the ideological father of capitalism, Adam Smith, in his *Wealth of Nations*, "so far as it is instituted for the security of property, is in reality instituted for the defense of the rich against the poor, or of those who have some property against those who have none at all." Inequality has always been the raison d'être of the police, to both preserve it and protect people from its consequences. Their role is both paradoxical and self-serving: to create crime (by defining its meaning in practical terms) while also preventing it. For this reason, the establishment of the Marine Police Office was a milestone in both "metropolitan policing and in the history of the wage-form." In this moment, we can see the theory of class division under capitalism put into practice: collaboration between property owners to direct the power of the civil state for their common interest against the poor.

In 1829, Home Secretary Robert Peel established the Metropolitan Police. His eponymous "Bobbies," patrolling the streets of London, were the climax of a process begun over three decades earlier by Harriot, Colquhoun and Bentham. Peel himself had learned the importance of preserving social order during his time trying to manage the colonial occupation of Ireland, and he therefore understood the value of a professionalized force. The model proved useful in many other sites of urban enterprise, including the newly formed United States of America, where immigration and industrialization were creating social and political havoc. The elite of America faced familiar problems as a settler-colonial state reliant on slavery. American capitalism needed organized civil institutions capable of clearing the land of its original inhabitants to then be tilled by slaves. Newly formed police forces proved crucial in the execution of the task. The scholar Alex Vitale summarizes it neatly: "The origins and functions of the police are intimately tied to the management of inequalities of race and class."

Why is it important to remember this? The history of policing helps us understand some key aspects of statecraft in the digital age. Like their historical counterparts, modern police understand the power of surveillance in enforcing a form of discipline. Safety on the London docks was defined as preserving social order and disempowering the lower classes, rather than reducing harm. Our current form of civil government similarly classifies citizens as an innately unruly mob, to be blamed for their miseries and disciplined into accepting the laws of the market. Law enforcement and intelligence agencies are incorporating technology into their arsenal in ways that would be familiar to those working the docks under the watch of the newly appointed police. As budgets for these agencies continuously grow, out of all proportion to the risks to public safety they supposedly guard against, this history helps explain the phenomenon.

And this tells us something about the idea of public safety, how we understand it and who defines it. By allowing law enforcement and intelligence agencies to tell us what it is to be safe, we allow them to use technology in increasingly oppressive ways. It also means we ignore the potential for technology to truly reduce harm and violence. Digital technology creates the capacity to improve public security and serve the public good, but not if we allow it to be abused by an authoritarian state.

We know about the sheer enormity of the American surveillance state because of Edward Snowden. In 2013, he leaked documents that showed how digital surveillance is a cheap and effective method of keeping tabs on local and foreign populations. PRISM, for example, which gathered metadata from an array of major technology companies, cost the relatively tiny sum of $20 million annually. Yet the capacity of the National Security Agency verges on the incomprehensible: in 2012, it was processing more than 20 billion communication events from around the globe (both on the telephone and the Internet) every day. The approach was summarized in one slide from an internal presentation: "Collect it All; Process it All; Exploit it All; Partner it All; Sniff it All; Know it All." It aspired to total informational awareness.

The monstrous scale of the ambition and practices of the NSA can make it difficult to make sense of its collective purpose. The agency struggled to analyze the tide of information it gathered every day; it encountered problems in merely storing it. A key challenge was that this kind of surveillance created a lot of noise but not many signals. But the Snowden documents also demonstrate that, in spite of these practical hurdles, there remained a coherent and long-term understanding of how it all fit together. In a candid presentation, an officer identified three factors that motivate the NSA and its surveillance programs: "National Interests, Money and Egos." American capitalism gained influence and profits from dominating the Internet during its early stages of development. This kind of technological project—total surveillance of communication—was part of an effort by the United States to "maintain its grip on the world," as journalist Glenn Greenwald put it. It recalls Marx's description of the modern state as "a committee for managing the common affairs of the whole bourgeoisie."

The NSA and their fellow intelligence partners in the Five Eyes (the name given to the intelligence alliance between the United States, Canada, the UK, Australia and New Zealand) were less than thrilled about the Snowden revelations, but a tacit understanding by the general public of the existence of this capability has had useful consequences for law enforcement and intelligence agencies. Multiple studies confirm that the idea of government surveillance profoundly impacts how we conduct ourselves in digital spaces, narrowing discussions online and reducing engagement with reading material perceived as contentious. Our liberties become conditional on their good graces. It is as though we are being disciplined into a new way of living, where any experimentation or whim, any deviation or mistake, exists under a threat of force. Old ways of working, communicating or engaging socially take on a new hue of fear, a watermark of pervasive surveillance. It is not unlike, we might imagine, how it felt to be unloading cargo from the Thames as part of the suffocating new category of waged laborer, under the watchful eye of the river police.

This misappropriation of power is not, however, some overzealous aberration. The NSA's surveillance capacity was relatively low-cost and efficient but only because it was predicated on specific kinds of political cooperation. Paul Ohm talks about how digital technology is creating a "database of ruin," where databases will eventually reach a point where they connect every individual to at least one closely guarded secret. This database of ruin is being built not by agencies themselves but by technology capitalism. Intelligence and law enforcement agencies ride on the coattails of the work done by companies to map our abstract identities, discussed in the preceding chapter. British police have paid data brokers to help them profile convicted criminals and estimate their chance of recidivism, and American police have used data from ancestry websites to identify people suspected of a crime. The Snowden documents show how state surveillance programs like PRISM functioned by tapping into the data flows generated by private businesses. The state obtains access to our personal spaces because technology capitalism has already beaten a path through our privacy defenses. Every web platform we participate in, every detail shared on social media, every item we are sold online generates data that can be accessed by the state, whether through legal processes or less formal ones.

It is not just that these companies inculcate a culture of sharing personal information (or just taking it without meaningful consent). They also rarely design data storage systems in ways that might protect customers by limiting the data they collect, deleting it, or notifying people when it is being stored. This means that they hand it over to state authorities with little, if any, reluctance. It can be hard to know where the companies end and the surveillance state begins; they share a common interest in sustaining the capitalism that overlays their relationship with users and customers. The outcome is that our digital lives are structured in ways that grant the state ever-greater powers, with ever-less accountability.

State surveillance is more heavy-handed and disciplinary, while surveillance capitalism is designed to appear consensual, convenient, diverting and mostly inconspicuous. But together they work in a complementary manner. Operating hand in glove, government and companies have created a technological ecosystem of multifunctional, cooperative surveillance.

The scale of the surveillance carried out by the Five Eyes is enormous, but it is not unprecedented. In fact, the social component of surveillance is appreciated more directly by the Chinese surveillance state than by its American counterpart. Far from trying to conceal how citizens are watched, the "Police Cloud" is a public program of information-sharing at provincial level that feeds into a national database run by the Chinese Ministry of Public Security. It uses records from people's medical history, supermarket memberships, deliveries, and other information linked to each person's national identification number. It also uses facial recognition software to identify suspected criminals. Researchers are developing a police car with a roof-mounted camera able to scan in all directions for individuals wanted by authorities. A jaywalker caught on one of the many cameras in the country (by 2020 there will be nearly half a billion) can be instantly identified as a repeat offender. These are not programs that are concealed from the public. Quite the contrary: the faces of jaywalkers, for example, are displayed on billboards above intersections in real time.

This disciplinary mode of surveillance operates in tandem with more social iterations. In 2014, the Chinese state announced its intention to implement a mandatory social credit scoring system by 2020. A trial is already underway. It is not clear which data might be used for the purposes of calculating each individual "citizen score," but reports suggest a score could be sensitive to particular purchases (buying diapers, for example, rates well, but alcohol does not) and opinions of the regime expressed online, even by one's friends (critical views will lower the score). A citizen's score affects them in myriad ways, including access to credit but also travel and various free or discounted services. For the state, the social credit system represents "an important basis for comprehensively implementing the scientific development view and building a harmonious Socialist society." Those who score poorly are socially isolated, confined to the digital underclass.

Many outside observers look at this naked pursuit of total, "harmonious" social control with horror, but the similarities between the Chinese and American states are difficult to ignore. The NSA may have done it covertly, and the Ministry of Public Security may do it more openly, but the respective agencies share ambitions and outcomes. State surveillance is not simply designed to detect wrongdoing; it is about creating a more general culture of compliance and isolating resistance. It is about generating social consequences for deviant behavior. It is a version of Colquhoun's "strong and overawing hand of power" as an effective complement to the locks and bars of a prison cell. And just as Colquhoun realized, modern states are conscious that digital surveillance is a highly efficient method of enforcing a social order. Digital technology enables a scaled-up version of the pilot program that Colquhoun supervised on the Isle of Dogs over two centuries ago.

The sophisticated integration of databases by Chinese authorities represents a significant and alarming technological advance—an improved capacity to turn noise into signals, as compared to the bulk collection of the NSA. The NSA, in contrast, appears to operate on the basis that effective surveillance does not require us to think we are being watched all the time; it can be powerful by simply creating the anxiety of *not* knowing when we are being watched. It establishes an implied understanding that intimate information can be retrospectively pieced together at any time to tell a certain story about us. But joining these data points into a cohesive picture remains the most important job for state agencies, and at this the Chinese state excels.

In contrast, the American approach has outsourced much of this work to private enterprise, and there is a lot of money to be made in this marketplace. Numerous profitable companies provide services to law enforcement using digital technology, to fill gaps in intelligence and policing. American data mining companies already boast about their capacity to generate the equivalent of a Chinese social credit score: Zest Finance, for example, at one point used the motto "All data is credit data." Major technology companies, like Google, provide data mining and analysis technology to police departments and city governments, as well as US intelligence and military agencies. Amazon is working to place voice-activated technology (such as Alexa) at the center of policing in the UK, by providing services for collating, storing and broadcasting data.

Palantir Technologies is perhaps the most notorious such company and certainly one of the most profitable in Silicon Valley. It has been described as "a darling of the US law-enforcement and national-security establishment." Palantir is tight-lipped about what it does, but its products assist organizations to easily search and find patterns in diffuse and diverse datasets. Its biggest customers are government intelligence and law enforcement agencies, and private organizations interested in fraud detection. Among other things, it is helping Immigration and Customs Enforcement (ICE) develop methods for policing immigrants in ways that are more sophisticated than ever. It is building software to help ICE access data from a range of sources across government departments that include information on foreign students, family relationships, employment, immigration history, criminal records and home and work addresses.

Palantir is doing the work that has already been done in-house at the Ministry of Public Security in China, a kind of entrepreneurial approach to shaping the capacities of public authority in a time of technological change. It was funded in its early stages by the CIA's venture capital arm, and its other notable investor is Peter Thiel, who sits on the board. "The most valuable businesses of coming decades will be built by entrepreneurs who seek to empower people rather than try to make them obsolete," wrote Thiel in his book *Zero to One*. This anodyne-sounding platitude belies something darker. Thiel has made a lot of money by using technology to empower a specific class of people: those who believe that the purpose of civil government is to allow authoritarians to keep the masses in line. In reimagining techniques of social control, Thiel looks increasingly like a modern incarnation of Patrick Colquhoun, the original architect of modern policing. Just as Colquhoun pursued his thought experiments on eighteenth-century dock workers, Thiel imposes his visions on twenty-first-century netizens.

This approach to statecraft aims to diminish the potential of democracy through imposing order. "The NSA has the greatest surveillance capabilities that we've ever seen in history," said Snowden. He went on:

> Now, what they will argue, is that they don't use this for nefarious purposes against American citizens. In some ways, that's true but the real problem is that they're using these capabilities to make us vulnerable to them and then saying, "Well, I have a gun pointed at your head. I'm not going to pull the trigger. Trust me."

This capacity for surveillance, using digital technology, has created a significant concentration of power in the hands of government, not the people. Snowden refers to Americans, but the argument is true for all citizens. The concentration of power makes it harder to hold governments accountable and easier for them to persecute their enemies. It makes it harder to organize resistance to authoritarianism. It is a reminder that the Internet is not inherently or necessarily a democratic space free from government intervention.

"What made the Internet so appealing was precisely that it afforded the ability to speak and act anonymously, which is so vital to individual exploration," writes Glenn Greenwald, reflecting on the Snowden revelations. "Only when we believe that nobody else is watching us do we feel free—safe—to truly experiment, to test boundaries, to explore new ways of thinking and being, to explore what it means to be ourselves." Collective private spaces—where we are able to communicate and collaborate in conditions of genuine anonymity, secrecy or both—are essential for experimentation and creativity, for exploring what we think without the ominous suspicion that we might have to pay for it later. The specter of state surveillance haunts these online spaces and diminishes our sense of what is possible, both personally and philosophically.

Snowden's revelations elicited some lofty language from lawmakers in the United States about freedom and government overreach. However, even after moderate reforms, the fundamentally unequal power relation between citizen and government has remained largely intact. The NSA still retains the power to tap into data streams collected and managed by private companies. The disturbing picture of government excess revealed by Snowden was met with a forceful claim on the part of the state that surveillance keeps us safe from terrorism. Elected representatives proved to be unreliable allies when it came to confronting the power of the surveillance state. If we want them to curb that power, we will have to find new political arguments and strategies for holding them accountable.

That is not to say that Snowden's revelations about these programs were fruitless. The mainstream adoption of encryption as standard by major technology companies is one testament to the effectiveness of his actions. (Strong encryption is increasingly under threat for this reason.) This technological response to Snowden's revelations has, at least to some degree, empowered people in many ways to resist government snooping. But individual, technical measures to resist surveillance will always be an incomplete solution. As Seda Gürses, Arun Kundani and Joris van Hoboken have argued, while the Snowden revelations were "disastrous" for the reputation of technology capitalism, "the emphasis on the technical aspects of the surveillance problem may have made the situation more manageable for them." Understanding surveillance in purely technical terms allows companies to recast the problem in ways that obscure their complicity. It reflects an ideal that "solutions for societal problems can come from technical progress and sophistication"—an ideal that neatly dovetails with the objectives of technology capitalism.

We now know more than ever about how the state works to surveil us. This is a phenomenon commonly observed in self-proclaimed liberal democracies as well as in authoritarian regimes. The accepted wisdom that underpins this power dynamic is that compromises of privacy are necessary because they allow the state to guarantee our safety. The reality is altogether different. We need to scrutinize the way the state and capital work in partnership to discipline us by building bars not with steel but with silicon.

"Everything that can heighten in any degree the respectability of the office of *Constable*," wrote Colquhoun, "adds to the security of the State, and the safety of the life and property of every individual." Colquhoun saw it as part of his mission, in documenting the significance of the pilot program on the Thames, to highlight the importance of a corps of professionals. Only a properly paid and uniformed cohort of officers could command the respect needed to perform their job. "It cannot be sufficiently regretted that these useful constitutional officers, destined for the protection of the Public, have been ... so little regarded, so carelessly selected, and so ill supported and rewarded for the imminent risques [*sic*] which they run, and the services they perform in the execution of their duty." Respect for the police, according to Colquhoun, should arise from an acknowledgment of the public service they provide in keeping the public safe. The Thames River police were supervising the docks, but Colquhoun also sought to draw an explicit link between policing and public safety in general.

Some two centuries later, in 1996, the American sociologist David Garland observed that "one of the foundational myths of modern societies [is] the myth that the sovereign state is capable of providing security, law and order, and crime control within its territorial boundaries." According to Garland, we are taught to believe that police are an essential, inescapable element of a functioning society. Even if we may have concerns about how their power is used, or want it to be reformed, the accepted wisdom is that without law enforcement there would be chaos.

These assumptions about the importance of police have gained new power in the age of digital technology. Our daily lives under surveillance capitalism involve countless algorithmic judgements made about us, defining our abstract identity. This technological capability is also used by the state, to regulate both public safety and deviance. The state collects data about us and makes use of it in discriminatory ways, and this is reflected in how the police do their job.

One place where we can see these dynamics in play is the growing use of predictive policing algorithms. Academic experiments on this kind of technology in the United States have been running for a number of years, often in partnership with industry. The basic idea is that this software tries to predict where crimes are likely to occur using a range of data sets—from traffic to the weather—to create a model for crimes such as burglaries, theft and assaults. By deploying police officers to hot spot areas, the theory is, the number of crimes will fall. In its early stages, this has appeared to meet with some success: a study from UCLA has found that predictive policing algorithms have reduced crime in the field. On the back of this kind of research, the industry that provides the relevant software products is growing quickly, with a range of consultancies selling predictive policing technology, including, predictably, Palantir. The gloss of technological precision applied to the messy work of policing is appealing, as is the idea that such programs can lighten the workloads of police forces and relieve some of the pressure on their budgets.

It remains unclear exactly how these models work—they are proprietary and therefore secret—but we know that one of the primary data inputs fed into these algorithms is crime statistics. If a burglary happens in one house, statistically that may affect the neighboring houses' risk of also being burgled; if a person is arrested for selling drugs in a particular place, past crime data may suggest that the location is a hot spot for dealing.

The problem is that crime statistics do not reflect the crimes actually occurring; rather, they provide a picture of the state's response to crime. By relying on this data, the software has a tendency to target low-income communities and minority neighborhoods, even though health data and population models indicate that drug use is similar across race and income groups. Similar concerns exist with stop-and-frisk policies, which have documented biases, more specifically based on race, which are then fed into the algorithm. Facial recognition databases also contribute to the problem. Law enforcement agencies hold records on millions of people for this purpose, largely unregulated. There are serious, documented issues with the accuracy of facial recognition software, particularly for people who are not white. Given that minority communities are overpoliced already, they end up overrepresented in these databases. It entrenches the criminalization of nonwhite communities, importing this discrimination into our digital future.

These biased data sets and algorithms, when used in a law enforcement context, have significant consequences. They generate a feedback loop shaped by racism and institutionalize certain understandings of risk. Both of these serve to confirm and exacerbate discrimination through the oversampling people who are already discriminated against, generating even more biased data that justifies further discrimination. Rather than some objective pursuit of criminality, these algorithms lead to confirmation of a historical tendency on the part of police to focus on poor people and racial minorities. Public notions of who commits crimes and what neighborhoods are safe become distorted.

The data on which we train technology "uncritically ingests yesterday's mistakes," as James Bridle puts it, encoding the barbarism of the past into the future. It is a history that is carried forward uncritically. The perception that these algorithms are scientific and objective goes unchallenged. This gives them a powerful allure for both industry and the state. Industry understands that there is money to be made in the promise of enhancing policing technology; the state appreciates the distance it puts between law enforcement and allegations of bias.

Many technologists and data scientists prefer a blunt metaphor for this use of algorithms: garbage in, garbage out. The discriminatory social trends that have already been exposed in everyday policing will continue to be reproduced in the supposedly more objective and scientific methodology of computerized predictive policing. Over time we can expect more data sets to be input, to better refine the algorithms. But given the biased way in which these algorithms are built and deployed, this is unlikely to actually produce a safer society; it will just let the police work more efficiently and effectively, which is not the same thing.

If these programs were really about safety, the algorithms might look quite different. They might target white-collar crime, for example. In fact, that could be a more suitable use for the technology. Despite being a complex phenomenon that resists simple definitions, academic research suggests that white-collar crime often follows clear and predictable patterns, particularly around the industries in which it tends to occur. Yet this kind of criminality, despite being highly damaging and widespread, is not the focus of predictive policing.

The use of digital technology by police reveals how ideas of deviance and safety are not objectively determined but are socially produced. Much in the same way that the eighteenth-century Marine Police Office created specific ideas of criminality through its work to prevent crime, digital technology is finding new ways to criminalize classes of people who are already oppressed. How the police decide what is "criminal" involves an array of decisions and policies that have been influenced by discriminatory social practices, reflected and reproduced in technological tools.

We can also see this phenomenon in action, perhaps with even more intensity, in American counterterrorism policy. The risk of harm as a result of terrorism is vanishingly low for most Americans, but the amount spent on policing that threat is disproportionately high. In other words, what it means to be safe is not necessarily about reducing the most pressing threats of harm, but rather specific threats that are politically determined. For example, documents revealed by Snowden set out the Skynet program run by the NSA, which uses an algorithm to identify terrorists. This algorithm was developed using data about "known terrorists" and comparing it with a wide range of behavioral data taken from cell phone use. The algorithm's highest-rating target was a man named Ahmad Zaidan. The NSA documents prominently, and with apparent confidence, label Zaidan a member of Al Qa'ida.

But Zaidan is not a terrorist. At the time, he was the Al Jazeera bureau chief in Islamabad. While Zaidan may have met with known terrorism suspects, traveled with them and shared social networks, he did this as part of his job as a journalist. While Zaidan may be a perfect fit for the algorithm to identify terrorists, it is immediately obvious to any human that he did not belong in this category at all.

To outside observers, the material about Skynet looks absurd to the point of being comical. But Zaidan himself offered a more sober assessment: "The allegations against me put my life in clear and immediate danger, when we consider that many people have lost their lives as a result of such fake information." These programs have real-world consequences: thousands of people have been targeted by lethal drone strikes conducted by US agencies over the last two decades. It is arguable that some, if not many, were like Zaidan and posed no threat to the lives of innocent people. This practice of data modeling, connected by government agencies to our real-world identity, puts us all at risk of arbitrary violence by the state. This technology legitimates unaccountable use of weaponized robots under the cover of a public safety project. It may not be wholly intentional; the NSA can hardly have wanted the kind of false positive generated by Skynet (though it is important to remember that President Bush discussed targeting Al Jazeera militarily in 2005, and the US military has been accused of bombing its buildings deliberately). But the capacity to treat people in this way—as unpeople—reveals both the dehumanizing ideology that informs certain kinds of foreign policy project and the power of technology to achieve those ends. The objectification of people by algorithms is a problem that will be explored further in the next chapter.

The Skynet program shows how technology accelerates the power of law enforcement to use violence arbitrarily on the disenfranchised and powerless. The recurring theme is that state agencies that do policing work create a specific idea of criminality and then use digital technology to "protect" us based on the fear they have engendered. It would be a mistake to see Skynet simply as an example of sloppy data science. It is a tiny insight into the myriad and secretive ways the state uses violence, guided by technology, for social control.

I do not mean to deny the existence of real threats to people's safety, from individual acts of interpersonal violence to the political violence of terrorism. Rather, my argument is that digital technology is being used for a more specific purpose, namely to substantiate the state's claim to be the great provider of safety, even when it does not achieve this objective. It feeds on a culture of what Adam Greenfield calls "unreconstructed logical positivism": a belief that the world is perfectly knowable, and that, with the correct inputs and algorithms, technical systems can generate solutions to all collective human needs. It also relies on what James Bridle calls "automation bias," our trust of machines to generate trustworthy responses because the computational processes are too complex and opaque to allow us space for criticism. By generating technical solutions to the problem of crime, the myth of the state as the indispensable provider of safety becomes a logical, indisputable truth. It creates a category of people it classifies as criminals, which it then polices and trains us to fear. This kind of statecraft has broad implications: it downplays other threats to our safety, glosses over the social causes of crime, and discourages us from thinking about ways in which digital technology might contribute to minimizing genuine threats.

How can we change this situation? It will require us to redefine what it means to be safe, and to imagine alternatives to the police. A safe society is not one that strips away the social security net and uses increased numbers of police to manage the fallout. We need to reject the neoconservative perspective, as Alex Vitale puts it, "that sees all social problems as police problems" and appoints them to be the lead agency in dealing with social, economic and political problems, armed with cutting-edge technology. We cannot let technology be used to upgrade the socially destructive phenomenon of modern policing.

We also need legal requirements for transparency in the state's use of algorithms in public decision-making. Unless the logic and data inputs in predictive police algorithms are publicly available, fundamental rights within the criminal justice system will be eroded. Such transparency will allow courts to test the reliability and accuracy of such programs and ensure they avoid arbitrary interferences with private life. It is true that these systems can be mind-bogglingly complex, with so many data inputs and variables that discovering their biases and appreciating their nuance can verge on impossible. But this is not a novel problem, and it is manageable. It is possible to imagine a statutory list of assumptions and data inputs that can be incorporated into this technology, and a similar list of prohibitions. The insurance industry is already regulated in some places in this way, allowing some factors to affect pricing of products and prohibiting others from doing so. Public participation in designing such rules will be necessary and valuable. If the makers of these programs lack the transparency to allow public scrutiny, we should stop using the programs.

We also need greater transparency over intelligence agencies. Oversight by regulatory bodies, reporting requirements, and better protection for whistleblowers who expose wrongdoing are some obvious steps toward this objective. We expect that in societies ruled by law, authorities will obtain a warrant before entering our private homes, so why not also require this for our private online spaces? The right to a private domain, free from government interference, has always been a right that needed fighting for. A similar struggle will be required today.

But it is also worth thinking a little more radically about the possibilities of technology to reduce harm. We need to decouple algorithms used to map trends in harmful behavior from law enforcement. It is possible to imagine a world where this type of data analysis is used to inform government spending and social programs, for example. A good example of where to start might be the Ceasefire program, which aims to prevent interpersonal violence through social intervention:

> Under Ceasefire, police teamed up with community leaders to identify the young men most at risk of shooting someone or being shot, talked to them directly about the risks they faced, offered them support, and promised a tough crackdown on the groups that continued shooting. In Boston, the city that developed Ceasefire, the average monthly number of youth homicides dropped by 63 percent in the two years after it was launched.

Rather than focusing on restricting gun purchase and ownership, which has largely proven ineffective, or randomized stop-and-frisk practices, which confirm police biases, a program like Ceasefire uses social relationships and community authority to address harmful behavior. There are all sorts of ways in which data and thoughtful algorithms could inform social interventions and the provision of services to try to de-escalate interpersonal crime, resolving many issues without the need for police intervention at all.

It is important that this kind of community work does not become a justification for more surveillance. Ceasefire has, for example, been implemented in New Orleans in less desirable ways. In 2013, Palantir used the city as a testing ground for its intelligence products focused on predictive policing. The program was rolled out almost covertly, with minimal oversight by elected officials. It was reportedly designed to work in partnership with the Ceasefire program, to ensure that community engagement preceded law enforcement involvement. But the emphasis ended up on prosecutions much more than on community engagement, ultimately undermining the intervention work. "It's supposed to be ran [*sic*] by people like us instead of the city trying to dictate to us how this thing should look," said Robert Goodman, a New Orleans native who worked on the Ceasefire program. "As long as they're not putting resources into the hoods, nothing will change."

The ideas that underpin Ceasefire start in the right place, focusing on empowering local activists and resourcing them to intervene into cycles of violence as peers rather than enforcers. But these ideas will only be effective, and make effective use of technology, if they resist becoming a cover for increased police power. Good policy design, with community involvement and accountability and drastic limits on police powers, can minimize this risk. In these contexts, it becomes possible to reimagine an idea of safety built on community practices that make use of technology in ways that are designed to reduce social isolation and risk. Such social interventions undermine the very need for a specialized police force.

Digital technology can help make our societies safer—it can reduce crime and protect us from other harmful behavior. But such outcomes are by no means guaranteed if we hand over powerful tools, developed in the digital age, to the police. "We are developing an official criminology that fits our social and cultural configuration," Garland warned two decades ago, "one in which amorality, generalized insecurity and enforced exclusion are coming to prevail over the traditions of welfare-ism and social citizenship." In other words, Garland was observing how a culture of fear was beginning to take precedence over ideas of collective responsibility for social problems over two decades ago. He has been proven more correct than he would have wished.

Modern law enforcement and intelligence agencies are established and sophisticated well beyond the wildest imaginings of Harriott, Colquhoun and Bentham at the end of the eighteenth century. In their day, these men faced "infinite difficulties and discouragements" as they strove to get a professional police force up and running to suppress "the extensive and enormous evils" that plagued their society—and convince civil government that such an approach would work. But the practical application of their ideas—particularly in the digital age—has generated its own valency. Agencies charged with the protection of the public have ended up serving their own interests above all. When law enforcement and intelligence agencies are so powerful that they sit above organs of democratic governance, it is almost impossible to hold them accountable. This creates a situation where public safety is not simply neglected or selectively defined; it creates a situation where public safety is put at risk.

In May 2017, cyberattacks affected 200,000 computers in 150 countries. This came with enormous human consequences, affecting universities and health systems. The attacks exploited a zero-day vulnerability in Microsoft software. A zero-day vulnerability is like finding a secret gate into a walled garden: it gives you access to do things you should not be able to do, but only for as long as the point of access remains undetected. Zero-day vulnerabilities are not intentionally placed there by programmers—rather they represent weaknesses or holes in the coding. The name "zero-day" indicates that the problem is still undetected by the person running the program, that is, the vulnerability has been known about for zero days. There is a whole market for finding these vulnerabilities and fixing them, and software companies spend a lot of time looking for them and sending out updates to patch them. Other hackers do it for pay. In this case, the vulnerability allowed hackers to lock the user out of the machine and everything on its hard drive unless a ransom was paid.

It seems that whoever launched this attack, known as the WannaCry worm, exploited a zero-day vulnerability that Microsoft had become aware of a few weeks before through the NSA. The NSA had seemingly found this vulnerability at some point before the attacks, possibly five years earlier, though we know few specifics. Rather than disclosing it to Microsoft to fix, the NSA kept it secret, as part of its efforts to accumulate a kind of digital arsenal; so long as it remained unpatched, they could exploit it for their own intelligence purposes. The problem was that at some point it was apparently lost or stolen, forcing the NSA to tell Microsoft about its existence.

This was not the first time the NSA had been slow to share its knowledge of security problems with widely used software. The Heartbleed vulnerability was a bug in encryption software used by websites that allowed attackers to eavesdrop on communications and steal data directly from services and users, including by impersonating them. When it was discovered in 2014, there were estimates that up to two-thirds of the world's websites were exposed, leaving large amounts of confidential information available on the Internet and allowing attackers to easily steal this information without leaving a trace. It had been identified by the NSA two years earlier.

As Microsoft pointed out in the wake of WannaCry, it was the equivalent of a Tomahawk missile going missing. Or like the theft of the only remaining stockpiles of smallpox, which have been hoarded by Russia and the United States. Or someone stealing a nuclear warhead.

When he publicly discussed these cyberattacks, Snowden touched on how this kind of statecraft is predicated on the idea that the price we have to pay for security is our right to privacy. For safety's sake, we have to accept that police and spooks will know all about us. We have to accept that our software will be full of secret gates into our private gardens, because it serves the purposes of security agencies. The way these cyberattacks unfolded proved how false this bargain is:

> This has never been a conversation about privacy vs. security. Because privacy and security improve together ... they are actually tied to each other. When one is reduced, the other is reduced. Surveillance and privacy are the contradictory factors. When surveillance increases, privacy decreases.

The NSA's idea of safety is different from how everyday citizens might understand the concept. The NSA seeks to accumulate digital weaponry and surveillance power at the expense of our individual freedom and privacy.

Moreover, what is at risk is much bigger than individual freedom and privacy. In 2012, Bruce Schneier observed how we are "in the early years of a cyberwar arms race. It's expensive, it's destabilizing, and it threatens the very fabric of the Internet we use every day."

The accumulation of digital weaponry—technological tools that allow spying or hacking of systems or other violations of digital infrastructure—is often a goal that takes precedence over privacy and security for the surveillance state. The kind of cyber-militarism that Schneier alluded to has only become worse in subsequent years. As more of our personal lives, social services and public infrastructure become integrated with digital technology, power shifts toward institutions that can control these systems, either legitimately or through nefarious means.

Cyber-militarism—and the tussle for power it represents—does not actually make us safer, it exposes us to risks. For some sections of society, which rely on the state to serve and protect their interests, the risks are worth taking. Those sections are much the same as they were in the days of Harriott and Colquhoun: they are those with wealth. The consequences of any risks, when they materialize, are often borne by everyday citizens.

Imagine, for example, if there was a zero-day vulnerability identified in the software used in a certain make of autonomous car or a weapon or an interface for a public transit system or power grid. If the NSA had some intelligence advantage to gain by keeping this vulnerability secret, it is hard to imagine them trying to patch it. This practice leaves us all exposed—a risk that only becomes more widespread and complex as more software gets incorporated into everyday products and systems.

The state has a long history of spearheading technological development, marshaling public resources to keep ahead of national rivals and to support domestic private enterprise. This work also feeds back into its objective to preserve the status quo. It discourages the idea that alternatives are possible, by encouraging us to internalize the feeling of being watched. As the history of the police shows, public and private power are often tangled together in ways that are deep-seated and mutually reinforcing. Despite nominally representing the public, the state in many ways serves the interests of the propertied classes first, reflecting the history of the London docks during the industrial revolution, and the police who began to surveil them.

It is time to shift this paradigm and start to think about ways in which digital technology can make our society more secure, in ways that are complementary with the freedom that comes with privacy. Data and computing can be deployed in ways that address social harms and tackle the problems that lead to violence. Whistleblowers like Edward Snowden are as much a constant of history as the bullies and thugs they expose. They are the surveillance agency of the people; they are the undercover spies for the powerless, and there are always more of them. We should protect them when they step forward, listen to what they have to say, use what we learn to hold power accountable, and never presume that this will happen automatically. Their presence is not a sufficient check on those in power. Rather, our job is to carry on the work they have started.

What began on the London docks two centuries ago was an experiment in state-sanctioned force and suppression for the purpose of protecting private property, justified by reference to public safety. We need to redefine what it means to be safe, and reexamine the threats posed to our collective security by unaccountable organs of state power. We need to break the locks and bars created by state surveillance, before we find ourselves in a digital dystopia.

Nearly a century after the establishment of the Marine Police Office, the London docks experienced a social transformation of a different kind. In 1889, the dock workers went on strike over their miserable pay. Other laborers joined them, until 130,000 people shut down the vital arteries of the city, in what came to be known as the Great London Dock Strike. Those on strike faced serious obstacles, but their campaign survived on the support from fellow workers as far away as Australia. "The proverbial small spark has kindled a great fire which threatens to envelop the whole metropolis," declared a news report. As the threat of a general strike grew, the dock companies relented, and ultimately the workers won their demands. It represented a historic reclamation of their power as a class of wage-earners, achieved by withdrawing their labor. The strike became a turning point in British labor history, a moment in which a new imagining of the potential of workers' power was brought into relief and set the paradigm for a bold and broad approach to labor organizing over the next century.

More than a century on, the moment is ripe for another shift in the balance of power, away from those who surveil and seek to control us, and toward those who seek to rekindle a society built on solidarity.

4

Technology Is as Biased as Its Makers

Exploding Cars, Racist Algorithms, and Design Beholden to the Bottom Line

In the late spring of 1972, Lily Gray was driving her new Ford Pinto on a freeway in Los Angeles, and her thirteen-year-old neighbor, Richard Grimshaw, was in the passenger seat. The car stalled and was struck from behind at around 30 mph. The Pinto burst into flames, killing Gray and seriously injuring Grimshaw. He suffered permanent and disfiguring burns to his face and body, lost several fingers and required multiple surgeries.

Six years later, in Indiana, three teenaged girls died in a Ford Pinto that had been rammed from behind by a van. The body of the car reportedly collapsed "like an accordion," trapping them inside. The fuel tank ruptured and ignited into a fireball.

Both incidents were the subject of legal proceedings, which now bookend the history of one of the greatest scandals in American consumer history. The claim, made in these cases and most famously in an exposé in *Mother Jones* by Mike Dowie in 1977, was that Ford had shown a callous recklessness for the lives of its customers. The weakness in the design of the Pinto—which made it susceptible to fuel leaks and hence fires—was known to the company. So too were the potential solutions to the problem. This included a number of possible design alterations, one of which was the insertion of a plastic buffer between the bumper and the fuel tank that would have cost around a dollar. For a variety of reasons, related to costs and the absence of rigorous safety regulations, Ford mass-produced the Pinto without the buffer.

Most galling, Dowie documented through internal memos how at one point the company prepared a cost-benefit analysis of the design process. Burn injuries and burn deaths were assigned a price ($67,000 and $200,000 respectively), and these prices were measured against the costs of implementing various options that could have improved the safety of the Pinto. It turned out to be a monumental miscalculation, but, that aside, the morality of this approach was what captured the public's attention. "Ford knows the Pinto is a firetrap," Dowie wrote, "yet it has paid out millions to settle damage suits out of court, and it is prepared to spend millions more lobbying against safety standards."

It is hard to imagine these days, but half a century ago car crashes were generally blamed entirely on the driver, despite cars incorporating very few safety standards into their manufacture. Every problem was attributed to the responsibility of the person behind the wheel. The automotive industry lobbied hard to limit its responsibility for deaths on the road and treated safety as incompatible with selling cars. "Self-styled experts with radical and ill-conceived proposals," warned John F. Gordon in 1961, "[think] the only practical route to greater safety [is] federal regulation of vehicle design." Gordon was the president of General Motors, and he was speaking to the National Safety Congress. His skepticism was undisguised: "The suggestion that we abandon hope of teaching drivers to avoid traffic accidents and concentrate on designing cars that will make collisions harmless is a perplexing combination of defeatism and wishful thinking." His address was met with enthusiastic applause. The powerful Business Council backed the industry, as did many other leaders of American capitalism. W. B. Murphy, president of the Campbell Soup Company, was openly disdainful: "It's of the same order of the hula hoop," he said of the concern about car safety. 'Six months from now we will probably be on another kick."

This attitude was, in part, a product of the lax regulatory environment. The national regulator was understaffed and underfunded. Under the Nixon presidency, key positions in agencies remained unfilled. In the late 1960s, Dr. Thomas Malone, chairman of the National Motor Vehicle Safety Advisory Council, wrote to the secretary of the National Highway Traffic Safety Administration about the problem of underfunding: "The Federal funding is not commensurate with the size of the problem. The serious gap between authorized and appropriated funding has handicapped the forward motion of the program." But the political establishment was not interested in applying much pressure to one of the biggest industries fueling the postwar boom in America. It was easier to adopt the line that road safety was a matter of individual responsibility than to tackle the industry head-on.

For these reasons, people continued to die—despite the known technological possibilities for making cars and roads safer. The National Academy of Sciences described the dangers of car travel in 1966 as "a neglected epidemic of modern society" and "the nation's most important environmental health problem." It is not certain how many people lost their lives or were injured as a result of Pinto fuel tanks igniting: estimates vary from hundreds to thousands. But the scandal acted as a lightning rod, prompting regulators to rethink the accepted wisdom touted by the industry and consider whether the requirements they imposed on manufacturers were sufficient.

The accounting system used by Ford engineers to weigh death and serious injury against cost and marketability was callous and impossible to justify. This was a preventable disaster. It was also bigger than any single company. In the 1960s, GM's Corvair had similar design problems that affected the car's steering and resulted in over a hundred lawsuits. The tragedy did not begin with the manufacturing of the car or even when it failed in testing. The tragedy, according to lawyer and consumer activist Ralph Nader, "began with the conception and development of the Corvair by leading GM engineers." It was an industry-wide culture that failed to consider the impact of design on the end user, that deferred and outsourced moral responsibility to consumers. Just like the Pinto, the Corvair was a problem by and of design.

This is not an attempt to pin blame on evil engineers or designers. The people who made these cars were working in a specific corporate climate. Their organizations were led by ruthless executives. The leadership of companies like Ford and GM ignored safety concerns in competition with other companies that did the same. It was not even a problem confined to the auto industry; there have been many other similar scandals involving corporate indifference to the human consequences of poorly designed consumer products. These scandals are not aberrant; they occur in a context, and to avoid them happening again requires a political strategy to attack the logic that produces them.

Companies will always seek to outsource responsibility for the negative consequences of their products, and lawmakers will often let this happen. We cannot just assume that civil government will properly regulate the development of technology and the design of algorithms, we have to demand it. Such campaigns can involve journalists and consumer activists, as it did in the campaign for proper regulation of vehicle design to protect road users. It can also involve industry-wide organizing among designers and engineers themselves—to make responsible and ethical design a workplace principle, rather than something that can be ignored.

Legal processes can be cumbersome and slow—the compensation Richard Grimshaw received for his life-changing injuries was never going to be sufficient for such suffering. But his case also initiated a movement of transformation—a chance to rethink how we make rules collectively to mediate the relationship between users and designers. Laws that protect people from dangerous products have an important role to play in our society. They set standards for corporate behavior, they downgrade the profit motive, they force governments to fund regulators to monitor compliance, and they provide a framework for understanding the wider implications of organizing society around the market. The demand for a rigorous legal regime that holds companies accountable for dangerous products is not a cure-all, but it is a worthwhile aim. It has the potential to be a non-reformist reform—a way to win concessions from capitalist systems and open up discussions about how to design things better. If handled well, in partnership with activists, journalists and workers, legal reforms of this nature can transform corporate behavior and create an agenda for arresting some of the worst excesses of capitalism, helping us to imagine how to do things differently.

Today the development of digital technology is taking place in an environment of very limited regulation, comparable to that which produced the Pinto in the 1970s. Iterative design processes encourage experimentation on unwitting users, with little transparency about how design decisions are made or their consequences. Users are held responsible for design failures. We can see this play out in biased algorithms and in more general trends in the development of networked technology, which often appear indifferent to the human consequences of bad design decisions. We are gradually coming to know some of the problems that this creates, but often the discovery depends on the work of dedicated researchers and observers of the industry. We need to build a broader movement to demand stronger forms of accountability—to ensure that design processes come under democratic oversight and incorporate ethical considerations. We need to catch the next Pinto before it comes off the assembly line and harms whoever ends up trapped inside.

Professor Latanya Sweeney of Harvard University typed her name into Google; she was searching quickly for an old paper she had written. She was shocked to see an ad pop up with the headline "Latanya Sweeney—Arrested?" Sweeney does not have a criminal record. She clicked on the ad link and was taken to a company website selling access to public records. She paid the sum to access the material, which confirmed she had no criminal history. When her colleague Adam Tanner did a similar search, the same ad for a public records search company appeared, but without the inflammatory headline. Tanner is white; Sweeney is African American.

Sweeney decided to study the placement of these ads to see if there was a pattern. She did not expect her results to be definitive. But her research produced a clear outcome: "Ads suggesting arrest tend to appear with names associated with blacks and neutral ads or no ads tend to appear with names associated with whites, regardless of whether the company has an arrest record associated with the name."

In other words, a greater percentage of ads with the word "arrest" appeared in ad text for first names that are associated with black people than for "white" first names. The presence of an actual criminal record did not appear to be the deciding variable.

How did this come about? To explain requires some unpacking of the online ad business. Each time you click on a website, an instant auction for ad space takes place between companies competing for your attention. As we know, surveillance capitalism has all sorts of ways for determining how valuable you are to a marketer, to allow platforms to make accurate bids on your eyeball time. These companies know more about our habits than we do. They have a detailed picture of our abstract identity—a history of our sense of self, defined by and for consumption—and they are using this information to send us marketing messages at optimal moments. Left to their own devices, this creates a situation where new technologies reproduce real-world forms of oppression.

The options around ad spaces are tailored in more ways than one. Google allows companies to tailor not just which audience sees the ad but also the content of the ad itself. As Sweeney explains:

> Google understands that an advertiser may not know which ad copy will work best, so an advertiser may give multiple templates for the same search string and the "Google algorithm" learns over time which ad text gets the most clicks from viewers of the ad. It does this by assigning weights (or probabilities) based on the click history of each ad copy. At first all possible ad copies are weighted the same, they are all equally likely to produce a click. Over time, as people tend to click one version of ad text over others, the weights change, so the ad text getting the most clicks

> eventually displays more frequently. This approach aligns the financial interests of Google, as the ad deliverer, with the advertiser.

Because of the way the algorithm has been designed, the machine learns to associate African American names with criminality. Even if you personally do not click on the ad, you experience the consequences of the machine learning from what other users clicked, which restricts the choices displayed to all subsequent users.

One possible response to this is that the algorithm is neutral, that it is just a vehicle for the ad, automatically responding to how people use it. Algorithms aren't racist, people are racist. But the algorithm is built in a way that also confirms implicit biases that exist in the real world, and it does so over and over again. The assumption that African American people are less trustworthy than white people is a commonly held form of implicit bias. It has real-world implications in a whole range of ways, from the success of job applicants to the split-second decisions made by police when pointing their guns at people. In Sweeney's study, we see this attitude reproduced in the world of digital technology, intentionally or otherwise. This is not a mystery or an unfathomable outcome. Google is not entirely responsible for racism having an impact on automated advertising, but it cannot shirk responsibility for it. Ford was not solely responsible for hundreds, possibly thousands of cases of people being burned alive in their Pintos, but the court of public opinion rightly felt that it could have easily prevented them if it had designed the car differently.

Part of the reason that this ad ended up being racist is because the process of designing algorithms and training them on real-world data takes place with basically zero transparency. The inputs are secret, and there is no formal regulation properly adapted to these processes. The bad experiences of users like Sweeney do not show up in the cost-benefit analyses for these companies when selling ad spaces. There isn't even a formal way to complain about it. There is barely a way to know about it. Biased algorithms exert considerable and increasing influence over many aspects of our lives. So long as they remain hidden or unexamined, we are allowing all manner of dangerous and oppressive practices to become embedded in new technologies, as machines learn to absorb the implicit biases that exist in the real world.

The idea that none of this is Google's responsibility loses sway when we consider that the power is in its hands to know and act on this information. Google decision makers knew the content the advertiser—their paying customer—plans to use. They are best placed to know the potential problem and how it might manifest, because they designed the system. At the moment, they calculate that the costs of finding and fixing these problems is higher than the cost of ignoring them, which is borne by others. We must find ways to change that calculus.

Google executives should bear the responsibility for the outcomes generated by their technology when it does what it is designed to do. In this case, Google is providing a service that is doing what it is designed to do: monetize advertising most effectively. In other words, the importation of racism into digital technology and the failure to consider implicit bias in design processes is not a bug, it is a feature of technology capitalism.

For this reason, what is perhaps most contemptible of all is that there is no actual reason for this kind of design discrimination to happen. The web is a space where oppressive attitudes could be structurally minimized, acknowledged and dismantled. Not only could we have policies in place that forestall racist ad placement, we could also engineer better representations of diversity that actively minimize prejudice. We could anticipate implicit bias and figure out ways to neutralize its effects in advance; we could prevent companies from capitalizing on its existence. We could design and build digital infrastructure that helps socialize people against discriminatory implicit biases. Such a prospect raises all sorts of interesting questions about how it might work in practice, and we could collectively start to grapple with this task.

We could campaign for and draft legal regulation of the design and engineering processes, much as consumer rights advocates demanded that the federal government implement safety regulations for cars. We need to establish rules that prioritize the goal of dismantling oppression over that of monetizing the web. This is a precious opportunity. If we simply wait for these problems to present themselves, or address them piecemeal as they emerge, we will miss the iceberg for the tip. We will allow an industry to entrench itself and marshal its forces against transparency and accountability while holding users accountable. At present, we are reliant on people like Sweeney to discover that these problems exist—and she only did so by accident.

Data scientist Cathy O'Neil, in observing the combination of careless logic, lack of feedback, and substandard data inputs that characterizes many algorithms, calls them "weapons of math destruction." She writes about how they have a tendency to "blithely generate their own reality." The perceived neutrality of digital processes provides cover for the sloppy and divisive treatment of people, while outsourcing the management of a range of activities to insidious effect. "Managers assume that the scores are true enough to be useful, and the algorithm makes tough decisions easy," she writes. "They can fire employees and cut costs and blame their decisions on an objective number, whether it's accurate or not." Computerized processes for determining answers to complex questions in the age of big data create exciting and transformative possibilities, but they also bestow on bad governance and management an undue gloss of accuracy and neutrality.

Through the small but disconcerting porthole opened by Sweeney, we glimpse a vast ocean of activity. Algorithms are being used in all sorts of ways that can have profound impacts on people. One example is reliance on automated processes for filtering job applicants, which can be biased against people with a history of mental illness or those for whom English is a second language. Another is standardized testing for college admission. Admissions processes, particularly if they do not require standardized tests, might use a predicted score as a proxy based on the applicant's demographic characteristics, with no clear idea of the accuracy of that substitute.

Algorithms are also used for deciding on parole applications, which rely on forms filled out by caseworkers without any indication of how the responses impact the output of the algorithm. In a deeply offensive example, a Google photo application, which automatically sorts photos by subject matter, once labelled a photo of some black people as gorillas. Secretive, proprietary algorithms have a tendency to produce thinly veiled bias disguised as scientific logic. These problems are not mere blunders, much in the same way that the bad steering in Corvairs or lack of buffering for the Pinto fuel tanks were not simply unfortunate errors. They are symptomatic of a flawed design process.

These kinds of processes affect people in society differently. As O'Neil points out, machines are cheap and efficient; their decision-making is more likely to be imposed upon poor people. "The privileged," she observes, "are processed more by people, the masses by machines." And it is almost impossible for people who confront these machines to question or challenge their decisions, if they even know that they are being made. Walmart, for example, has created catalogs for low-income audiences, which market a disproportionate amount of junk food relative to healthier options. Data about arrests can also be used in oppressive ways when cross-referenced with other data sets. Evidence of an arrest can mean that automated résumé-sorting software may pre-emptively exclude a candidate from consideration for jobs or deny one access to consumer finance, sometimes even after the arrest has been expunged from the public record. Machine learning is routinely used and tested on the poor, and it is the most vulnerable in society that end up dealing with the consequences.

There is undeniably a class dynamic to the impact of oppressive algorithms. Technology, especially under the stewardship of the elite, mirrors the value systems that underpin social divisions. Many of our current conversations about the dangers of artificial intelligence are dominated by the possibility that such technology will lead to a third world war. However valid, the framing of these concerns reveals something deeper. Many of the people driving these conversations are rich white men—and, as researcher Kate Crawford points out, "perhaps for them, the biggest threat *is* the rise of an artificially intelligent apex predator. But for those who already face marginalization or bias, the threats are here."

There is a growing and sophisticated network of algorithms that generates all sorts of social, economic and cultural consequences. Abstract identification invariably relies on profiling based on data, or discrimination in relation to our abstract identities. This is a practice of *data discrimination*. That is, communities and individuals are segmented into audiences for marketing purposes, often on the basis of superficial assumptions about specific and incomplete data, which has highly divisive effects and accelerating impacts. As Adam Greenfield argues, "contemporary technologies never work as stand-alone, isolated, sovereign artifacts." Networks collect and exchange data, which has been channeled in various directions as a result of path dependency, and this is amplified by market functions and social prejudice. The decision-making function of machines has the capacity to not just reproduce traditional social fault lines but also to exacerbate them.

Consider payday lenders. Payday loans are mostly taken out by people who are struggling. In the United States, five groups have a higher than average probability of using them: those without a four-year college degree; people who rent; African Americans; people earning below $40,000 annually; and people who are separated or divorced. It is a highly predatory and destructive industry: over 80 percent of payday loans are renewed or rolled over within two weeks, and 22 percent of borrowers will repay more than they borrowed in fees before they manage to discharge the loan. Payday lenders find customers through a careful process of data collection, cross-referencing and curation, often involving third parties, who sell information about the potential customer to multiple companies to target for business. In essence, users of payday loans are locked into an abstract identity, which results in an onslaught of intrusive marketing. In response to lobbying efforts and a public campaign, Google's advertising division agreed to ban payday loan ads that meet certain criteria. This is a terrific first step but by no means a complete or adequate solution to the problem.

An even more poignant example is the for-profit education sector. Nearly every private college in the United States earns most of its revenue from multi-billion-dollar federal financial aid programs. They do this by luring students to enroll on the promise of social mobility. These universities charge up to 20 percent more than state and community college equivalents, yet more than half of students will withdraw four months later without graduating, saddled with debt that they will struggle to pay over the course of their lifetime. Like payday lenders, these colleges specifically target vulnerable people. A presentation revealed to a Senate committee in 2012 outlined tips for recruiters about the kinds of people who were likely to enroll: "Welfare Mom w/Kids. Pregnant Ladies. Recent Divorce. Low Self-Esteem. Low-Income Jobs. Experienced a Recent Death. Physically/Mentally Abused. Recent Incarceration. Drug Rehabilitation. Dead-End Jobs—No Future."

In other words, people with a deep-seated desire to overcome their current, often highly stigmatized predicament and desperately looking for a way to do so.

"A potential student's first click on a for-profit college website," writes O'Neil, "comes only after a vast industrial process has laid the groundwork." This includes tricks like finding the "pain point"—the moment at which the person is most open to taking drastic steps to improve their situation, often revealed through Google searches or college questionnaires. But it can also be found out by buying information from companies that post fake job ads, or ads promising to help obtain Medicaid or food stamps, and then asking callers if they are interested in college education. Poor people are basically stalked into enrolling, with colleges routinely spending more on marketing than tuition. It is hard to see how a harried single mother or survivor of abuse, desperate to grasp hold of any chance to improve her life, stands much of a chance against this kind of manipulation. And as O'Neil points out: "The for-profit colleges do not bother targeting rich students. They and their parents know too much."

These marketing practices segment us into particular groups in an effort to capture some of the most powerful forces in our psychology —shame, desire, guilt—and entrench and exacerbate these emotional states for the purpose of making money. They are computerized sale techniques made possible because of surveillance capitalism and manipulative algorithms that violate basic understandings of ethics.

The continual process of abstract identification privileges the powers of data scientists based on the collection of our data without our knowledge, using methods of analysis that we do not understand. It creates a digital existence in which decisions are made for us, with our worth measured in arbitrary, unaccountable ways. It facilitates a modern equivalent of redlining, where people are discriminated against on assumptions about the demographic qualities of the population. Indeed, for a time, Facebook explicitly allowed a form of digital redlining, giving marketers the option of excluding certain people from seeing their ads based on "ethnic affinities." They have also admitted to allowing job advertisers to exclude people of a certain age from the audience for specific ads. Both of these practices are probably unlawful, but—once again—it is almost impossible to detect them without information from the companies themselves. It requires laborious effort on the part of journalists or academics after the fact, when we should be trying to catch these problems and mistakes before the code is shipped.

By 1978, under pressure from advocates and the regulators, Ford agreed to voluntarily recall all Pintos built between 1971 and 1976. This decision came just months after a jury awarded $126 million in damages to Richard Grimshaw (the award was reduced by the trial judge but remained substantial). The decision in *Grimshaw* was affirmed on appeal, with the court noting that "the conduct of Ford's management was reprehensible in the extreme." It found that management "exhibited a conscious and callous disregard of public safety in order to maximize corporate profits ... endangering the lives of thousands of Pinto purchasers." Months later, criminal proceedings were brought against Ford for the deaths of the teenaged girls in Indiana. Ford was acquitted in this case. But it did end up settling subsequent claims against it made in relation to the Pinto. By 1980, the model was discontinued.

The Pinto scandal should not be seen as an aberration of amoral engineers who failed in the design process to properly value human life. While it is important for everyone involved in the design process to be thoughtful and aware about their work, it is also important to acknowledge that these engineers and designers were operating in a business-driven context. The executive leadership at Ford made critical decisions and ignored important information that was given to them. They also operated in a fiercely competitive market, where regulators were asleep at the wheel or, worse, captured by the industry. Changing this dynamic required the work of journalists, activists and lawyers. It also required new forms of regulation that anticipated the risk of dangerous design and created conditions in which engineers could proceed ethically without jeopardizing their employment. Under capitalism, it is an unceasing battle to defend human life and dignity from the magnetism of the bottom line.

Computer code itself functions as a form of law. It is written by humans and it regulates their behavior, like other systems of power distribution. It is not an objective process or force of nature. It expresses a power relation between coder and user, and it will reflect the system in which coders work. "Code is never found," Lawrence Lessig reminds us. "It is only ever made, and only ever made by us." Letting the free market determine these matters means that digital technology risks reproducing discrimination under the cover of an inscrutable process. Joy Buolamwini is a founder of the Algorithmic Justice League, which aims to publicize and challenge bias in algorithms: "We don't have to bring the structural inequalities of the past into the future we create," she argues. In her view, we can only achieve this if we organize around a specific purpose with intention.

Some legal prohibitions on discrimination already exist that would capture some of these examples as they manifest in biased code. But there are not enough such curbs, while enforcing them will require updated powers for regulators. Identifying these problems will also involve imposing greater duties on tech companies. We need to demand that lawmakers and public agencies, under color of democratic authority, intervene in these markets and both promulgate and enforce design requirements on the industry. "A democratic government is far better equipped to resolve competing interests and determine whatever is required [to improve transport safety] than are firms whose all-absorbing aim is higher and higher profits," wrote Nader in 1965 of the car industry. The same is true of technology companies and government today.

Importantly, workers who build this technology also have a role to play in changing the culture of design. Ethical design considerations can serve as an industrial and political organizing tool, acting as a bulwark against predatory business practices. "Technological professionals are the first, and last, lines of defense against the misuse of technology," writes Cherri M. Pancake, president of the world's largest organization of computer scientists and engineers, the Association for Computing Machinery. In 2018, the organization published an updated code of ethics, which requires developers to identify possible harmful side effects or the potential for misuse of their work, consider the needs of a diverse set of users, and take special care to avoid the disenfranchisement of groups of people. Among the feedback received about the code by the ACM was this comment from a young programmer: "Now I know what to tell my boss if he asks me to do something like that again." Resolving the ethical considerations involved in design is rarely a straightforward task, but it is not impossible, and creating space for technologists to consider the options and work through them fully is an important component of this.

We can start to see how this looks in practice already, as workers in major technology companies are taking on their bosses on ethical grounds. Microsoft employees organized to demand that their company cancel its contracts with Immigration and Customs Enforcement and other clients who directly enable them. "As the people who build the technologies that Microsoft profits from, we refuse to be complicit," they wrote. "We are part of a growing movement, [comprising] many across the industry who recognize the grave responsibility that those creating powerful technology have to ensure what they build is used for good, and not for harm." This ethical and ultimately political decision to refuse to build harmful technology is not made on an individual basis, but a collective, industrial one. Similar insurgencies are happening at Google, with 4,000 workers signing a petition opposing a project on behalf of the military and senior engineers refusing to work on specific projects that would enable the company to win sensitive military contracts. This kind of collective organizing has immense potential to transform the culture of technology production through self-organization, which opens up ethical questions much more effectively than any top-down form of discipline or compliance.

Twenty-first-century technology has the capacity to throw off prejudice and treat people with dignity, rather than reinforce traditional notions of privilege. But this future requires that we make visible and address the architecture of digital technology that is reproducing discrimination and oppression due to the monetization of the Internet and the commodification of its users. We need to find ways to actively implant principles of nondiscrimination into how digital technology is structured, to ensure that those accessing it are doing so on equal terms, not being palmed off to be processed by machines. An essential part of achieving this will be a culture in which makers of digital technology are conscious of, and consciously reminded of, the impact of their actions.

Design processes in the digital age need to make it easy for engineers to take more account of users' interests. But how we come to understand users' interests can be a complex question, and it will take time and effort to incorporate these interests into design processes. As we develop technology and explore its potential, we may also have to impose limits on our technological capabilities if we are to avoid creating harm.

This is a particularly important consideration as we witness the expansion of the Internet of Things—as more everyday objects are fitted with network connectivity. You can buy a fridge or oven or a home climate system that is connected to the Internet and can therefore transfer data between objects and people. The range of products under development (which often span from excessive to useless, it must be said), also reveals the potential positives of this kind of technology, such as the assistance it could offer to people with mobility problems in the home. Amazing progress is being made with technology built to help with various forms of disability. There is also the promise of convenience: if your smart suitcase fails to show up, you can go online and track it down.

But there is a troubling side to smart objects too. As we bring more devices into our home that communicate with outsiders in ways we cannot control, the Internet of Things is arguably turning into an enormous surveillance apparatus. This is a particular problem for the vulnerable. As Elise Thomas writes, on the topic of technology and domestic violence:

> Advances in technology are both a blessing and a curse for the targets of domestic violence. New "smart" technology can make it easier for them to contact help and document the abuse—but equally, it can be misused to monitor their activities, eavesdrop on their conversations and even track their location in real time …
>
> Once upon a time, a phone number could be enough to get someone killed—so what does it mean for the targets of domestic violence when we move into a world where it's possible for someone with access to the right devices to track every move, hear every breath, read every heartbeat on a screen?

The implications of the Internet of Things for survivors of domestic violence are profound. As more devices become connected to the network without our ability to control these data flows, it becomes much easier for others to access large amounts of information about us. Wearable technology can be hacked, cars and phones can be tracked, and data from a thermostat can reveal whether someone is at home.

This depth and breadth of data is frightening for anyone who has experienced an abusive relationship—that is, for a lot of people. More than a third of women and more than a quarter of men in the United States have experienced rape, physical violence, or stalking by an intimate partner in their lifetime. Technological abuse is now standard practice among people who choose to use violence. In a 2014 survey of service providers for survivors of domestic violence, 97 percent report that their users experience harassment, monitoring, and threats by abusers through the misuse of technology. This is often harassment and abuse by phone, such as text messages and social media postings. But 60 percent of service providers also reported that abusers have spied or eavesdropped on the children and the survivor by means of technology. Abusers do this by giving gifts to the child or planting devices on the child's belongings; 11 percent reported instances of toys with hidden "spying" technology. The survey also found that 45 percent of programs report instances of abusers trying to locate survivors through technology. These findings were supported by another piece of research that found that 85 percent of the shelters surveyed work with survivors whose abusers tracked them using GPS, and 75 percent work with survivors whose abusers eavesdropped on their conversation remotely using hidden smartphone apps. Nearly half of the shelters surveyed ban the use of Facebook because of fears about revealing locational information to stalkers.

These social problems are not the responsibility of tech companies to fix, but they are an undeniable feature of the society in which companies operate. They are something that ought to be considered in the early development stages and accommodated during the design process. We are often told how handy and futuristic it is to connect more and more personal devices to the Internet. But not everyone feels that way. The generation of huge quantities of personal data, and our inability to control how this data is collected and stored, has serious consequences, especially for certain groups. Yet the experiences of large sectors of society, particularly those that are vulnerable, commonly appear to be absent from the design process.

This approach ends up affecting everyone, not just those with specific vulnerabilities: as technology capitalism finds new ways to learn about our personal lives, we can expect the government's spies to be finding their own way onto this information gravy train. In testimony submitted to the US Senate in February 2016, James Clapper, then director of National Intelligence, conveyed this very well. "In the future, intelligence services might use the [Internet of Things] for identification, surveillance, monitoring, location tracking, and targeting for recruitment, or to gain access to networks or user credentials," Clapper said. The state often makes use of industry innovations to repurpose them for their own interests. Writer Evgeny Morozov put it succinctly: "In case you are wondering what 'smart'—as in 'smart city' or 'smart home'—means: Surveillance Marketed As Revolutionary Technology."

This indifference on the part of companies to the experiences of their users stems in part from a specific understanding of usability and utility. The automotive industry was in a similar mind-set when the Pinto scandal broke. In *Unsafe at Any Speed*, Ralph Nader points out that it was largely uninterested in spending money to improve safety. About $166 was spent on research for every traffic fatality, a quarter of which was from industry. By way of comparison, industry and government together spent $53,000 on safety work arising from each death in the aviation industry. While car companies were happy to invest in making cars faster or slicker, when safety features threatened aesthetic design principles, the industry resisted them. Technology companies, insistent on connecting everything to the Internet of Things and designing products with a certain kind of user in mind, occupy a similar paradigm. They like to talk about servicing customers, but this purpose is understood through a specific and narrow frame.

One of the most obvious reasons for this is that the people designing much of our digital technology are drawn from a specific demographic. The overrepresentation of white men in Silicon

Valley is well known. According to Reveal, from the Center for Investigative Reporting, in 2016 ten large technology companies in Silicon Valley did not employ a single black woman. Three had no black employees at all and six did not have a single female executive.

Some companies are doing better than others, and many major companies release diversity reports, which is a change from years past. But white people, particularly men, remain overrepresented in the tech industry both relative to the population and compared to the private sector overall. This trend is even more extreme among executive leadership. It is for this reason that we see design approaches to networked devices that are indifferent to threats like domestic violence, despite the fact that it is a prevalent problem within the community. Given a specific cohort of people in the room, the decisions made by these companies will inevitably display particular biases, and there will be a lower chance of anyone thinking to correct for them.

There is a class dynamic to this also. Adam Greenfield points out that the Internet of Things is being designed by a group of people who have completely assimilated services like Uber, Airbnb and Venmo into their lives, despite the fact that this does not reflect a universal experience. They have embraced the digitalized, individualized, optimized and commodified version of the world: "These propositions become normal to them, and so become normalized for everyone else as well." In reality, a significant number of people have never used these services or even heard of them. But they are not the constituency being serviced by technological development. One journalist observed that the 2018 Consumer Electronics Show seemed "less about real innovation breakthroughs solving unmet needs and more about incrementally improved nice-to-haves for the 1%." Another commentator was more blunt: "[San Francisco] tech culture is focused on solving one problem: What is my mother no longer doing for me?" Too often, the people involved in the design process come from an experience of affluence (as well as a particular gender), and this affects the development of technology more generally, with repercussions for us all.

Diversity within the ranks of programmers is vital to changing this culture. This is not just a pipeline problem but also a problem within technology companies that will require changing things like recruitment practices, accountability processes, and policies relating to working conditions. The Google walkout of 2018 involved 20,000 employees of the company stopping work to protest how the company manages sexual misconduct cases and highlight the implications for women in the workplace. The workers won some of their demands almost immediately, but there is more to be done. It serves as an inspiring example of the sophisticated barriers that exist to cultivating a diverse workforce and how they might be dismantled through organizing. Tech companies ignore this at their peril. The lack of diversity in the industry is now so glaring a problem that calls for change, and proposals for achieving it, attract widespread and mainstream attention. While the prospects of success in this respect are an interesting topic for reflection, they are already the focus of considerable discussion and activity. I want to turn to a broader question.

Changing the representation of diverse groups within the maker class is not enough. We need to change the culture around ethical design. Executives who urge their programmers to move fast and break things clearly expect someone else to pick up the pieces. We need to argue that the paradigm should instead be about building thoughtful programs that are respectful of the impact of design and thinking critically about the identity of the user. Ethical quandaries should not be considered above the pay grade of the programmer, they ought not to be mentally outsourced to others—but that requires programmers to have the skills and the capacity to navigate them. This will mean expanding existing educational programs on ethics and making them more mainstream. But to put such programs into practice, tech companies must also provide the space for accommodating these deliberative processes. That may also mean prioritizing human involvement in decision-making and moderation over automated processes, even when it is more costly and less efficient.

Creating a programming culture that places greater value on the importance of empowering people and limiting the risk of harm is a necessary (if insufficient) step to addressing some of these problems. Over time, such work might broaden into questions about political power and ultimately foster a culture that celebrates technology for peaceful purposes and challenges its prevalence in industries of violence and oppression, such as prisons, policing and the military.

People buy and use the available products voluntarily, of course, and we respect their right to make these choices. But it is impossible to think that every one of these people fully understands the nature and implications of recent technology. Samsung caused controversy when it included in its Smart TV privacy policy a warning to customers that they should "be aware that if your spoken words include personal or other sensitive information, that information will be among the data captured and transmitted to a third party through your use of Voice Recognition." Even a Barbie is not safe from Internet connectivity: Mattel released a version of the doll that uses Wi-Fi to send data back to the company for research and development. But she comes with her own vulnerabilities. Security researcher Matt Jakubowski reported being able to hack the doll, despite significant steps by the manufacturer to protect customer privacy. "It's just a matter of time until we are able to replace her servers with ours, and have her say anything we want," he said.

We are witnesses to an insatiable drive to connect everything to the network, both to sell new products and to use the data generated to give companies a competitive edge. Many companies overlook the safety implications of this development. These are not just outcomes of poor attention to risk or security planning, though it would be helpful if such concerns were not routinely an afterthought. The problem is more deep-rooted: it is that this drive for technological development is motivated by capitalist enterprise. The companies making smart objects have a business model: they want to sell us things, and they want to collect more information about us to sell to other companies that wish to sell us things. Their motivation is not to develop the best or most useful or safest piece of technology we can imagine, it is to flog the most profitable.

Profiteering and entrepreneurship may have brought us good and useful advances in technology, but it would be a mistake to dismiss the intensely negative consequences as merely the growing pains of our digital existence, and it would be even more egregious to imagine that these advances somehow represent the fulfilment of the potential of digital technology. The sad reality is that these developments in technology are not particularly revolutionary: they represent the continued direction of human ingenuity toward the purpose of making money, the same old expansion of the market into yet another frontier. Technology capitalism shares similarities with fledgling industries of the past that became rapacious. They have significant potential to improve our lives, but only if we manage to hold them accountable to something other than the bottom line.

A culture of ethical programming that respects diverse perspectives will help address some of the problems that are emerging with biased technology. But we also need democratic oversight over this maker class. In his analysis of the automobile industry in the 1960s, Ralph Nader argued that secrecy was one of the policies most inimical to the improvement of car safety. "Not only does this industry secrecy impede the search for knowledge to save lives … but it shields the automobile makers from being called to account for what they are doing or not doing." We can see a similar force at work today when algorithms are used as a substitute for human decision-making without proper transparency or accountability. Proprietary algorithms used by the government are often kept secret for security reasons, or by the manufacturing companies for commercial reasons, so they can charge for the use of their product. This lacuna of transparency and accountability undermines equality of opportunity and conceals inequality in outcomes. We need to force "black box" algorithms open.

Algorithms designed to analyze DNA evidence for use in criminal cases is a good case in point. DNA evidence is becoming a highly complex field, as testing for the presence of DNA can occur with increasingly tiny samples. As a result, samples almost always display DNA mixing, where multiple matches can be made to individuals. This can happen with objects touched by different people over hours or even days. The extent to which each person contributes to the mix depends on many factors, such as the rate at which they shed DNA material, rather than simply the order in which they touched it. Such complex samples are becoming difficult to analyze, especially by human lab technicians, and have produced results that were later discredited. In such a climate, government officials are increasingly relying on computerized processes to analyze these samples, processes often provided by private companies.

Without transparency over how these programs work, there is a gross potential for injustice. DNA evidence is compelling to juries, and a computerized process for generating results only strengthens this tendency. In New York City, some defense lawyers objected to the use of such evidence on grounds that the method was not accepted as reliable by the scientific community. The defense lawyers were denied access to the code for the program and so were unable to ascertain the logical inputs that went into the algorithm. With the help of science and math interns and forensic experts, legal aid lawyers managed to reverse-engineer the software. This was a monumental effort from public defenders, and one that ultimately bore fruit. After listening to extensive expert testimony, the judge decided that such evidence was unreliable, hence inadmissible. But this ruling was not the end for computerized DNA testing; it is still used as evidence in jurisdictions across the United States.

If public decisions are made about a person—especially if they involve a person's liberty—there should be a right to know how this decision was reached. We no longer convict people in secretive courts, without laws of evidence, because it is rightly considered unfair. Lawyers subject expert testimony to rigorous cross-examination, and are careful to assess whether the credentials of the witness are sufficient to support the conclusions they draw. Justice must be seen to be done. In this context, DNA evidence produced by black box algorithms can be highly influential and also scientifically flawed. A more transparent process for building these algorithms is essential to protect against faulty logic working its way into our system of justice. Computer programs ought to be treated like expert witnesses. We ought to subject a program's assumptions to a similar level of scrutiny, not treat it like an inert provider of objectively determined truth.

There is no reason why such programs could not be developed by a public authority or using public money, or could not be subject to an audit by such a body. There could be reporting guidelines, a certification process for determining whether programs produce biased outcomes, or any number of other regulatory regimes. LRMix Studio offers an alternative: it is an open source software product that interprets complex forensic DNA profiles. Similar open source tools have also been developed for matching samples to DNA databases in a way that reduces the risk of false positives and negatives. To reach the standard of reliability for acceptance by the scientific community, especially on an ongoing basis, this kind of transparency is essential.

There is a danger, of course, that making these formulas transparent gives people the opportunity to subvert them. Dedicated criminals might learn how to avoid shedding DNA on crime scenes, for example. But these problems are hardly novel, they do not foreclose other methods of evidence collection, and they do not form a good enough excuse to entrench a world of baked-in bias. The privilege against self-incrimination and the right to legal counsel are accepted as vital to a functioning criminal justice system. They both also happen to make it easier for guilty people to walk free. Nonetheless, we accept them as necessary for the proper administration of justice. The same ought to be true for computer programs used to aid decision-making in the criminal justice system.

Much like how we expect transparency over budgetary decisions or public resource allocation, algorithms used for public decision-making purposes should be available for scrutiny to ensure that the logic and data inputs are fair. There are growing calls for governments to reject the use of black box algorithms in public decision-making processes and open up all code to public scrutiny. These are good starting points and can be part of a longer-term objective to apply this kind of scrutiny beyond just public bodies.

Private companies are now at the center of research into the sophisticated algorithms that go into machine learning. The academy is no longer able to compete with the vast resources that have been centralized by the data boom in companies like Google, Facebook and Amazon. This is particularly true as the technology industry invests heavily in machine learning, sucking its skilled professionals into private enterprise, largely excluding the public from the benefits of these developments. According to the vice president of Microsoft Research, Peter Lee, the cost of acquiring a top researcher in 2017 was around the same as the cost of signing a quarterback in the NFL. As *Wired* observed, "since then, the market for talent has only gotten hotter ... big players are now buying up AI startups before they get off the ground."

The basics of this philosophical field—figuring out how to teach machines to imitate human intelligence—will eventually become more readily available to other organizations over time as they are deployed. But there are further ways in which the development of this technology is held hostage by a small collection of companies. This kind of research requires extensive computing power and large amounts of data, both of which tend to be concentrated in large tech platforms.

The logical inputs that inform machine learning are deeply complex, to the point that even individual engineers might struggle to explain how a machine comes up with a particular response. In such circumstances, rigorous testing and standards are essential to catch problems before they are unleashed on the public. While there have been important recent initiatives oriented toward self-regulation of the industry, these are not sufficient. We need intervention by public, democratically accountable authorities. We need to start thinking about how we can infuse machine learning, like all fields of technological development, with principles of justice and fairness and make them accessible so that the benefits of these advances can be shared. We need to understand how concentrations of power are standing in the way of this objective.

In Dowie's documentation of the Pinto disaster, he discusses all the ways in which Ford and the relevant regulator, National Highway Traffic Safety (NHTS), negotiated the drafting and implementation of new standards for the industry. Dowie outlines how Ford was able to delay, for years, the implementation of standards that it would struggle to meet without a costly redesign. The NHTS did eventually find that the Pinto contained a safety defect, and this prompted its recall by Ford. But subsequent analysis of the scandal reveals how the safety defect found by the NHTS was the result of manipulation of the standard testing. Among other things, the NHTS increased the speed of crash tests; it used a different kind of vehicle as the "bullet car" in the collision with the Pinto, weighted to make maximum contact with the fuel tank; and it ensured that the headlights were switched on to provide a possible source of ignition. The NHTS justified these changes on the basis that they would bring the tests closer to real-world scenarios. But the clear effect of this redesign was that the car would fail. The Pinto burst into flames upon impact.

The implication was that these modifications to the standard testing process were done in response to public outrage, generated by Dowie's article and the lawsuits. Ford was held to standards that other companies were not—standards it had no way of knowing that it was required to meet. Ford has never admitted that it did anything wrong.

Yet the lesson is this: as we learn more about an industry and what is possible technologically, we need to update our expectations in terms of safety and accountability. We need to organize activists, lawyers and journalists to highlight the human consequences of badly designed technology, and force the industry to adapt to design culture that values safety and works to mitigate bias. We need to demand that governments intervene in this industry to establish publicly determined standards and methods for holding companies accountable when they are breached. The standards must be constantly updated and responsive to changing circumstances as we learn more about the problems and experiment with solutions. Just like we would not let a car onto the road without crash tests, the parameters of which are subject to public scrutiny and influence, algorithms and products should not be inflicted on the public if they have not met certain standards or been tested for certain biases before shipping. We need to create a feedback loop for good design that allows lessons learned in the field to inform improvements to a product.

Technology capitalism is very much alive to this potential. Silicon Valley is commonly associated with libertarianism, but the politics of the industry is more complex. Surveys of elites in the technology industry show that they in fact share a distinct set of values that correspond with liberal views and policies, and this means that they are commonly associated with progressive values. But tech elites differ from the traditional liberal base in one significant way, which has important implications for a number of issues raised in this book: regulation. The tech elite is positively inclined toward the laws of the market and entrepreneurship. They understand that this is the key to their success. They reject the idea that they should be regulated, putting them at odds with progressive calls for government intervention.

Why shouldn't we have crash testing for artificial intelligence? Or a certification process for machine learning? Why shouldn't we have a panel of experts who are required to remain independent from industry, who can offer guidance on testing and addressing bias before code is shipped or security risks before a product is sold, and investigate instances when they managed to slip through the net? We need to demand broader and more representative samples of people that beta-test new technologies, to ensure designers get feedback from outside the average pool of users. We need to incorporate proper feedback loops and meaningful appeal channels for people who are subject to automated decision-making, so that mistakes are not left for users to resolve individually. We must find ways to avoid privileging existing data over the relatively unknown, to avoid importing biases from available data sets and underestimating the limits of our knowledge. We need more humans supervising automated decisions, and we must stop unthinkingly using the latter as a replacement for the former. We need to develop public guidelines about best-practice algorithms and design processes, and empower agencies to monitor standards in ways that are accountable.

"The regulation of the automobile must go through three stages," wrote Nader. "The stage of public awareness and demand for action, the stage of legislation, and the stage of continuing administration." We ought to apply a similar approach to technology capitalism today, in which we scrutinize the industry, rally around demands for change, and put in place processes for ongoing accountability. Such regulatory processes are hardly perfect; they can be cumbersome, subject to industry capture and misdirection. But that is equally true for any number of industries that are vital to our health, such as food safety monitoring, and the regulation of medical products, and the automotive industry. Algorithms, technological devices and artificial intelligence, when badly designed, expose us to risks as much as an unhygienic meal or faulty pacemaker. If used by government, they might even compromise administrative processes that govern our social security or judicial decisions about our personal liberty. We need to think about them as products that are designed and can be modified, and reject the arguments put forward by the tech elites to deny responsibility for their impact and blame users instead.

Lee Iacocca became president of Ford at the very moment the company was deliberating whether to enter the small car market. He convinced other executives to rush the Pinto into production. The Pinto was known as "Lee's car," and Lee's corporate leadership was essential to its development and the catastrophic decision-making that underpinned it. Yet, nearly half a century later, Iacocca has gotten off virtually scot-free from his involvement in the scandal. Rather than serve any jail time for the horrendous damage caused by the Pinto, Iacocca went on to have a well-regarded career in the industry. He even considered running for president at one point, proving that creating dangerous products is no barrier to aspiring to the highest political office.

We live in an era when certain makers of technology are also perceived as presidential material. We should heed the lessons of the Pinto scandal: companies will exploit the absence of regulation to make products that do not properly account for the human consequences of bad design. Unless we find ways to hold them accountable, this will not be considered a mark of shame but an emblem of success. Far fewer people die on the road nowadays than they did during Iacocca's time, and we can learn from this example to achieve similar improvements in the safety and fairness of algorithmic design.

5

Technological Utopianism Is Dangerous

The Tech Billionaires Have Nothing on the Paris Commune

Julian West lay down to sleep in his Boston home on the evening of May 31, 1887. He customarily slept in a sealed chamber, a soundproof cellar below his house, in an effort to address his chronic insomnia. To assist him in falling asleep, West also hired the services of an alternative therapist. "I called in Dr. Pillsbury, [who] called himself a 'Professor of Animal Magnetism,'" explained West. "I don't think he knew anything about medicine, but he was certainly a remarkable mesmerist."

He was indeed. As a result of a fire that destroyed West's house that night, and a particularly profound form of mesmerizing animal magnetism, the insomniac awoke on the afternoon of September 10, 2000—113 years later. He was discovered by the new owner of his house while excavating the back garden, prior to building a subterranean laboratory. The new owner explained to West that he was now in what had been the distant future.

West remained unconvinced of this passage of time, and it was only when his host showed him the skyline of modern Boston that he began to realize his predicament. "Every quarter contained large open squares filled with trees, among which statues glistened and fountains flashed in the late afternoon sun. Public buildings of a colossal size and an architectural grandeur unparalleled in my day raised their stately piles on every side," he recalled. "I knew then that I had been told the truth concerning the prodigious thing which had befallen me." West spent the remainder of his life exploring this wonderland of a future, with its advanced industrial organization and domestic automation ensuring a bountiful existence for all citizens, the product of technological development.

This may seem like an overly elaborate set-up to a novel, but it was one of the most popular fictions of its day. Published in 1887, Edward Bellamy's *Looking Backward* was arguably "the most widely read and influential utopian tract of all time" and inspired "Bellamy clubs" across the country. The idea of a utopian future in which the problems of the present were miraculously resolved and society was organized efficiently and fairly captured the fancy of a whole generation.

Imagining a better world has occupied humans for millennia. It is a powerful elixir, the idea of a society of stability and sufficiency; it can be a source of motivation to strive for something better, as well as a method for highlighting the shortcomings of the present. Utopianism is the practice of taking these imaginings to extremes, applying innovative and creative ways of thinking to work out the logistics of a future perfect society. But problems arise because these visions of the future are often detached in causal terms from the troubles of the present. Utopians prefer taking imaginary shortcuts to a new world, like Bellamy's protagonist's overly long nap, offering little guidance about how to actually get there. The Boston in which West awakens feels vigorous and modern, a society motivated by equality and respect, powered by futuristic technology. Though Bellamy never identified it as such, it was commonly understood as a form of utopian socialism. But while we are free to imagine this perfect society, we remain fettered by its impossibility. If this kind of thinking creates a yardstick by which to measure our progress and an ideal to aim toward, it also renders our imaginations a salve rather than a cure.

Bellamy's novel was exciting because it reframed basic assumptions about the intractability of human misery. "It is a mystery," muses the host from modern Boston, Dr. Leete, reflecting on the past West used to inhabit, "how men with children could favor a system under which they were rewarded beyond those less endowed with bodily strength or mental power." Bellamy's future was one of universal sharing and industriousness, and such images have a moral weight. But the method of arriving at this destination remained persistently obscure throughout the novel. "All that society had to do was to recognize and cooperate with [industrial] evolution," explains Dr. Leete, with little further detail. Technological optimization, it seems, was Bellamy's pathway to utopia. The assumption has had a long legacy.

There is another tendency running through some forms of utopianism that can also serve the darker purpose of obfuscation. The idea that these imaginary future societies are actually detached from the present is often, ironically, an illusion. Without a proper understanding of the nature of the problems of the present, these visions for the future offer solutions that look remarkably similar in structure to the world they are trying to escape. Utopians often seek to beguile the problem of politics, to bypass negotiating the messy and complex networks of power relations between ruler and ruled. But in doing so, it is too easy to fall into the trap of reproducing the structures that give rise to current misery, with little more than a vacuous overlay of supposedly reasonable, enlightened thinking by those with already enough power to appoint themselves to the job of shaping such societies. Utopianism can abridge our capacity to imagine what true emancipation might look like.

In *Looking Backward*, for example, Bellamy paints a futuristic society made up of a "federal system of autonomous nations," each with its own "industrial army," that looks not unlike a benevolent version of state-run, monopolistic capitalism. It is never clear how this social and economic system avoids the problems that West had left behind in the previous century. If we fail to transform the structures that give rise to social and economic inequality, advances in all fields of human endeavor will remain beholden to these problems—and so too will the imagined solutions.

The challenge is to use our imaginations without falling into undiscerning, insidious utopianism. Can we come up with an ideal society in our thoughts but one that also exists with reference to the present? Right around the time Bellamy was writing, Oscar Wilde made a forceful case against capitalism in his essay *The Soul of Man under Socialism* and expounded the vital importance of a world in which "poverty was impossible." In it, he was right to argue that "a map of the world that does not include Utopia is not worth even glancing at." But it is also worth remembering that a map of the world that includes Utopia must be grounded in some kind of material reality; it will offer a route for getting there.

Technological utopianism is a specific vision of the future. It sees technological progress as the means to bring about a perfect society. To this extent, technology serves as a remedy for human weakness, doing a range of jobs faster and better than we ever could, overcoming the limitations of the natural environment, and creating the foundations of a genuine meritocracy free from the chaos of human society. Its supporters believe that technology, developed to the extreme, or optimized, has the capacity to generate a virtuous society of abundance.

Utopianism has a long history. But the specific kind of thinking popularized in Bellamy's *Looking Backward* saw its heyday in America between the 1880s and the early 1930s. Technological utopians were living in times of labour unrest, arising from the grimy and at times brutal developments and after-effects of the industrial revolution. Their work illuminated some of the anxieties of their day and applied constructive thinking to how technology could address them. Their movement, insofar as it could be called one, shared certain values and tendencies. From a theoretical perspective, it seized upon what they saw as the most important contemporary trend in society—the development of technology—and attempted to predict the outcomes of its advance and spread.

The shared vision stitched together from these individual texts reveals several common threads. These men (as they all were) had exciting hopes for the future; there is something precious about their optimistic exercises in fantasy.

The result is a set of imaginary worlds that look remarkably similar: highly organized, free from the filth and chaos of the industrial revolution, clean, efficient, and able to satisfy the needs of its inhabitants. Technology was shown to improve every aspect of human life—from production to housework, urban planning, and even the weather. How exactly this would happen was never fully explained, but it tended to involve large-scale machinery designed by engineers and tended to by workers. New inventions, such as personalized, compact electric motors, or glass with the strength and malleability of metal, facilitated the construction of vast and beautiful cityscapes and urban living spaces. Communal kitchens produced meals for everyone, punctually, without waste. Education was widely accessible; dangerous work was obsolete. The end of selfishness and monopolistic enterprise reduced inefficiencies and raised the living standards of the traditionally disenfranchised.

But while this was imaginative and innovative, there was also something of a conservative strain running through the ideas of technological utopians. Historian Howard P. Segal talks of how their writings often formed "a movement not of revolt but of antithesis —a movement seeking to alter the speed with which American society was moving but not the direction." They saw themselves not as creators of something new but as advocates for a progress already underway. Their writings resisted a sophisticated examination of the deficiencies of the present by conjuring a distinct and better future.

This left an unavoidable black hole in the logic of first-wave technological utopians. They argued that a future perfect society was a logical extension of the present—it could be generated by animating existing social trends—yet it was unclear how this might come about in practice. Very little is ever said about this point: the onset of utopia was typically skipped over as simply a logical outcome of technological advancement. To the extent the point was addressed, utopian writers described their ideal societies transpiring as a consequence of various vague factors that accelerated development, such as greater intelligence among the masses, recognition of and cooperation with industrial evolution, the unstoppable expansion of corporations, and the transformative ability to manufacture power. In other words, technological utopia was understood as the natural progression of capitalist development.

Glossing over the intricacies of cause and effect is a convenient storytelling device, but it also reveals something deeper about this kind of thinking. It manifests a desire to avoid a proper reckoning with the causes of problems in the present and, as such, any clear theory about how they might be changed. For some technological utopians, like Bellamy, this took the form of a mostly benign escapism, which provided a rich and compelling portrait of an equitable world of human flourishing. While there was diversity in how they approached their narrative, the technological utopians writing in the decades after the industrial revolution displayed common tendencies and philosophical bents, and they sparked a movement of sorts focused on a brighter future. But whether unconsciously or by design, many of these utopian worlds had a way of echoing trends of inequality that existed in their own society. It was symptomatic of a certain form of political blindness.

Most obviously, this can be found in the set of social and political relations that complemented the utopian concept of economic order. For example, in many of the technological utopians' writings, those with a mastery of technology formed a new elite class, commanding respect and deference. In George S. Morison's book, *The New Epoch as Developed by the Manufacture of Power* (1903), engineers were literally revered:

> [the engineer] is the priest of material development, of the work which enables other men to enjoy the fruits of the great sources of power in nature, and of the power of mind over matter. He is the priest of the new epoch, a priest without superstitions.

The rest of society was obliged to toil on the machines these engineers created. In King Camp Gillette's manifesto *World Corporation*

(1924), humans became part of the machinery itself—"cogs in the machine, acting in response to the will of a corporate mind." Just as the capitalists dictated to workers in the industrial revolution, so too would engineers preside over a society that treated people as a kind of machinery.

This instrumental understanding of people entailed markedly ascetic views about free time. In *The Diothas* by John Macnie (1883), for example, people only worked up to three hours a day, yet inactivity was reproved: "If any live in idleness, it is evident that others must toil to support them. Time-honored custom, therefore, requires that all children, whether boys or girls, shall acquire some handicraft." In *Looking Backward*, neglect of work in the "industrial army" was punished with bread and water in solitary confinement. Similarly, despite labor-saving technology that meant that people need only work two hours a day, it was not a life of leisure in Henry Olerich's *A Cityless and Countryless World* (1893). "We teach that labour is necessary and honourable, that idleness is robbery and disgrace," declares the protagonist, a visitor to Earth from a utopian society on Mars. Workers were expected to devote any free time to self-improvement. Free time in utopia, it seems, was not actually free. This kind of attitude fits with the prevailing views at the turn of the century, when industriousness was highly valued and respected. (Indeed, Bertrand Russell penned an entire dissenting essay in 1932 called *In Praise of Idleness*, affirming that "there is far too much work done in the world, that immense harm is caused by the belief that work is virtuous.")

Given the obsession with efficiency and productivity as the path to self-actualization, self-control was the defining feature of intimate relationships in these technological utopias. Social interactions were definitively distant. Reproduction was often considered the sole purpose of sex, with licentiousness considered "unchaste" and "impure." Though basic forms of equality were recognized for women—in ways often far ahead of their time, and even ahead of our current state of affairs—family structures and sexual mores remained rigid. Though some technological utopians envisaged lives of fierce social and political equality for women, the usual whiff of sexism often accompanied such visions. "It is their privilege," remarks the guide in *The Diothas* in reference to women, "to be beautiful, and in some measure, a social obligation to keep themselves so." Their job, in that particular utopia, was mainly housework. The reproduction of discriminatory attitudes to women in these imaginary societies would come as no surprise to many modern women, given it was mostly privileged men penning the words.

Technological utopians also cast off democracy; technical expertise was much more important for running a virtuous society. "Politics which we recognize as a necessary governmental part of our competitive, industrial system will have no place under the 'World Corporation,'" wrote Gillette. "There will be no voting, no political campaigns, and no favourites of fortune, either socially or industrially, except those who by study, application, perseverance, intelligence, and ability earn and by right attain positions in the World Corporate System." Voting was seen as a poor way to make decisions, especially when technology and technocrats could do it better. "Administration in a technocracy has to do with material factors which are subject to measurement," argued Harold Loeb in *Life in a Technocracy* (1933). "Therefore, popular voting can be largely dispensed with. It is stupid deciding an issue by a vote or opinion when a yardstick can be used." According to these writers, the optimization of technology would free society from the problem of politics, and social democracy would be superseded by more logical and cohesive organizing principles.

These utopias often put on display an imagined solution to unequal economic distribution and social disharmony that was just more of the same: more capitalism, technologically optimized. The entire premise of Gillette's utopia was that society be run as a giant corporation. He believed the industrial and political system was similar to a piece of machinery, which could be improved with the help of a scientific approach to design. To "complete the industrial evolution" wrote Gillette, the idea of a singular, enormous corporation "has power to join in harmony millions and millions of individuals, and, where chaos reigned, order and system takes its place." The democratic organization of society stood in the way of advances being made by the corporate structure: "The most serious obstruction to material progress is our present Government and its political parties and machinery ... These same men would have been among the enraged mob who destroyed the looms in an English mill less than a hundred years ago."

The description of the "logical evolution" that society underwent in Bellamy's *Looking Backward* was almost identical to Gillette's idea of a single corporation:

> The nation ... organized as the one great business corporation in which all other corporations were absorbed; it became the one capitalist in the place of all other capitalists, the sole employer, the final monopoly in which all previous and lesser monopolies were swallowed up, a monopoly in the profits and economies of which all citizens shared.

In *A Cityless and Countryless World*, such monopolies are shunned, but utopia is still achieved via an extreme form of capitalism: "*free* competition, of *healthy* supply and demand" rather than substandard "monopolistic laws."

While these technological utopians showed dazzling imagination and an infectious enthusiasm for the future, they were also susceptible to some surprisingly conservative and unsavory ideas. In some cases, these intellectual shortcomings stemmed from an indifference to the social and political work that needed to be done to transform their visions into reality. In other cases, they derived from an uncritical embrace of capitalist economic and social relations, which inevitably influenced how technology developed. Often it was a mix of both. As capitalism escalates, and especially as technology develops, political, social and economic democracy necessarily decreases. This was a criticism that many of the technological utopians were unconcerned by, or unwilling to entertain. The result was that these various utopias were either appealing but unachievable, or achievable but not very appealing.

Today these texts feel clunky and outdated. The movement, insofar as their shared vision and common popularity constituted one, dissipated. Segal argues that World War Two turned the idea of a utopia arising from technological development into a dystopia. There was a drift away from optimistic faith in technology after the Vietnam War; as the environmental damage associated with technology became clearer, the economy lost momentum and conservatism became ascendant throughout the seventies and eighties. As technofixes to political problems became ever more obvious attempts to paper over complex and deep-rooted social divisions, the public's belief in the virtues of technology faded. An associated skepticism toward expertise set in, which was accentuated when coupled with ideas of wasteful big government (in the United States, at least). Technological progress became more commonly associated with dystopias. Technological utopianism was a product of its time, and people were no longer buying it.

Why bother engaging with these writers from a forgotten movement? We have not realized the dreams of the first wave of technological utopians, that goes without saying. But in retrospect, these thinkers displayed an ironically unintended prescience, notably with regard to the social implications of the widespread technological development that ended up happening well after their lifetimes. Remarkably, technology capitalism is once more upheld as a significant contributor to improving our lot, to overcoming the problems we face in the twenty-first century. In the digital age, we are witnessing an ideological revitalization of the idea that utopia can be achieved by building more extreme versions of technology capitalism. It is a revival that imperils the ideals of democracy and equality.

The creators of the comedy television series *Silicon Valley* make regular visits to the headquarters of various tech companies to gather material for the show. They describe the Valley as a place where the hippie culture of San Francisco has "run headlong into rampant capitalism." One of the main characters in the series is Gavin Belson, the slick and ruthless CEO of a Google-like company, Hooli. In one of his many pep talks to staff, he declares: "I don't know about you people, but I don't want to live in a world where someone else makes the world a better place better than we do."

Technology capitalism, from the late nineties onward, has become a redeemer of American capitalism at large. The exposed brick and steel of the proverbial Silicon Valley office seems like a world away from the decline of the manufacturing industries of the 1980s. Companies that design and sell technology are now some of the world's most profitable corporations. The tech industry has revived the economy, as part of the shiny project to bring together markets and communities all over the world.

Moreover, they are harbingers of convenience and interconnectedness in our individual lives, a picture of upbeat, self-actualizing capitalism. Eric Schmidt and Jonathan Rosenberg enthuse about how "infinite data and infinite computing power create an amazing playground for the world's smart creatives to solve big problems." They praise technology, like Google's smart contact lenses, that "continuously tracks a person's vital signs." Bernard Harcourt has written about how the Apple Watch allows us to measure, track and improve ourselves: extracting deeply intimate information, it creates a unique profile of intimate life. The technology is positioned as objectively futuristic: it can detect early symptoms of a heart attack, hours before ordinary medicine could detect anything. But it can also be tailored to the wearer: it gives us "the most complete picture" of our movements, so we can customize our exercise goals. This mass-produced piece of technology "makes the subject flourish in his or her individuality." Surveillance capitalism, disguised as a drive for personal efficiency, becomes equated with self-actualization.

This is the world that Richard Barbrook and Andy Cameron portrayed in their influential essay "The Californian Ideology," published in 1995. The authors described a burgeoning political philosophy centered on achieving individual freedom "through the natural laws of technological progress and the free market." Californian ideologues offer "technological proof to a libertarian political philosophy … therefore foreclosing on alternative futures." Gavin Belson may be a caricature, but he is not so distant from many real-world figures in Silicon Valley who "promiscuously [combine] the free-wheeling spirit of the hippies and the entrepreneurial zeal of the yuppies." Detecting a mix of utopianism and fatalism, Barbrook and Cameron posit that this way of thinking sees political disputes about the alternatives as essentially meaningless. The future has arrived, even if it is yet to be evenly distributed outside the pleasant climes of Silicon Valley.

The social consequences of these developments read as eerily familiar. Engineers and tech entrepreneurs are treated with quasi-religious awe. It would not be a stretch to describe Steve Jobs, for example, as wielding priest-like power when he presided over the minimalist temple of the Apple Store. Elon Musk has been profiled as having "the boyish air of a nascent superhero [who] says his ultimate aim is to save humanity." Or take *Time* magazine on Mark Zuckerberg: "He's extremely smart, but he doesn't have any of the neurotic self-consciousness or self-doubt that often accompany high intelligence. His psyche, like his boyish face, is unlined … His faith in himself and what he's doing is total."

It would be too easy to go on. Suffice to say there is a reverence for engineers and entrepreneurs who have presided over the rise of digital technology. They are looked to for guidance, for solace, for solutions to the problems plaguing humanity.

In other interesting ways, the utopians' vision has come to pass: the tech gurus, as before, are almost exclusively white men. The rigid gender roles that still persist in digital spaces echo some of the descriptions offered in first-wave technological utopian literature. Despite decades of progress for women's liberation in the twentieth century, the insular world of technology sometimes seems immune, as though hermetically sealed off from the outside world. Social media continues to be a cesspool of misogynistic abuse, and contradictory approaches to representations of women and their treatment prevail online. It all reflects a highly nineteenth-century attitude to gender equality. This is perhaps unsurprising, given that women are still grossly underrepresented in the technology industry: in the United States, women make up a quarter of the tech workforce, while in the UK it is 17 percent.

To take one example, consider Facebook's policy on posts involving nudity. This policy has resulted in the repeated removal of images of breastfeeding; the deletion of an image of a fat woman in a bikini advertising an event about feminism and fatness; censorship of photos of Australian Aboriginal women performing a public ceremony as part of a protest, bare-breasted and wearing traditional body paint; and censorship of the iconic Pulitzer Prize–winning photograph by Nick Ut of children fleeing a napalm attack during the Vietnam War, which includes the naked, nine-year-old Kim Phúc. Meanwhile, images of nude and semi-nude women in sexualized and objectified poses are permitted to remain on the site in enormous volumes, especially if they depict celebrities. Implicit in these so-called community standards is an undeniable political agenda about women: sexualization and objectification of women's naked bodies is appropriate for the Facebook "community" (whoever that might be), but politicized images are less so. We are often told that traditional sexual roles and moralistic judgements of the choices we make in our personal lives are passé in the twenty-first century, but digital technology seems to have developed with these trends baked in.

Similarly, the technological utopians' ideas about politics have been reincarnated in recent decades. Social democratic parties in representative democracies increasingly consider themselves not as power brokers, ideologues or even representatives but rather as technocratic managers. Regulatory initiatives should be light-touch, according to this received wisdom, to avoid hampering "disruptors"—people who displace old ways of thinking with transformative new ideas. "Regulations get created in anticipation of problems," warn Eric Schmidt and Jonathan Rosenberg in their management book for Silicon Valley, *How Google Works*, "but if you build a system that anticipates everything, there's no room to innovate."

Neoliberal democratic governments have been outsourcing state responsibilities to the private sector for decades now, and often see themselves not as agents for redistributing power but as experts tinkering with the large machinery of capitalism. Large-scale welfare programs, like health and education, are constantly being rationalized and often outsourced to the private sector in the name of running them more efficiently. President Barack Obama assessed his first term as one characterized by "the disease of being policy wonks," describing how he was "very comfortable with a technocratic approach to government." More recently, the response to President Donald Trump's right-wing populism has been to devalue democracy and put more faith in elites as managers of social and political affairs. In the lead-up to the 2016 election, an influential essay by Andrew Sullivan entitled "Democracies End When They Are Too Democratic" argued for more control by elite political managers, "to protect this precious democracy from its own destabilizing excesses." While many of the technological utopians' projections did not come true, it is significant that the recent march of digital technology has coincided with an obsession with the perceived benefits of technocratic management of economic policy and other state functions, at the expense of democratic processes.

If anything, this desire to replace politics with technology has gone further in the twenty-first century, in ways that might surprise even the original technological utopians. Mainstream, legitimate players in the tech industry are seriously discussing and building whole new societies.

Balaji Srinivasan, an investor, entrepreneur and academic, has famously outlined what he calls the "ultimate exit." This may sound like a scheme to board the Hale-Bopp comet, and the reality is only a little less drastic. Srinivasan was referencing the economic theory about voice versus exit. This is the idea that members of a group (such as workers in a company or consumers in a market) have two choices when they are dissatisfied with deteriorating quality of the benefits of being a member: they can use their voice—by complaining, for example; or they can exit the group altogether. Srinivasan reckons that it is time people did the latter: "build an opt-in society, ultimately outside the US, run by technology."

Elon Musk, the so-called wunderkind turned tech entrepreneur, started SpaceX in 2002 with the ultimate aim of colonizing Mars. He plans to launch the first cargo mission in coming years and recruit settlers to send out within the decade. Describing the spirit of these missions, Musk equates it, unabashed, with "the settlement of the New World by the colonists who crossed the Atlantic Ocean centuries ago." In his portrait of the digital age and its future, *Lo and Behold: Reveries of the Connected World*, Werner Herzog interviews Musk about his plan to settle humans on Mars; Musk claims it is "in case something goes wrong" with Earth. In response, the astronomer Lucianne Walkowicz points out in the next scene that "the focus on making Mars habitable distracts us from the fact that we are currently making Earth uninhabitable." With his widely admired company SpaceX, it seems that Musk is trying to use technology to literally create a whole new world as a way of exiting the current one, and in doing so he manifests a total acceptance of the settler-colonial narrative. His plan is to buy and exploit his way out of a dysfunctional society with his own self-designed utopia.

Peter Thiel, the venture capitalist, had focused his advocacy on sea-steading. In a 2009 essay, Thiel famously sets out how he "no longer believe[s] that freedom and democracy are compatible." He yearns for "an escape from politics in all its forms" by abandoning nationhood and settling the seas. Interestingly, Thiel includes a call to resist technological utopianism, at the same time as advocating for a form of utopianism:

> The future of technology is not pre-determined, and we must resist the temptation of technological utopianism—the notion that technology has a momentum or will of its own, that it will guarantee a more free future, and therefore that we can ignore the terrible arc of the political in our world.

Ironically, Thiel has seemingly since abandoned his plans on the basis that they were perhaps unrealistic, conceding in 2015: "I'm not exactly sure that I'm going to succeed in building a libertarian utopia any time soon." He might be better off engaging a mesmerist and waiting out his perfect libertarian future, like Julian West. But he ought to prepare himself for a pretty long sleep.

There is a difference between technological determinism, which is perhaps a better description of what Thiel outlines above, and utopianism. It is possible to be an optimist about the future of technology and humanity, without that bubbling over into fantasy. Optimistic imaginings are important and even necessary. But the problems arise when we start imagining or even trying to create perfect worlds by taking current trends in technological development to extremes. This is not to say that we should stop using our imaginations or stop getting excited about the potential for technology to improve our lives. But how this technology is developed, and the extent of the improvement, are ultimately political questions. We should not confuse the advancement of technology in its current form with technological advancement for society as a whole.

Perhaps most importantly, technological utopianism blinds us to the ways in which it is possible to build a new world from political materials available in the present. We do not need an apocalypse or an exodus of the elite while the world burns in order to save humanity from itself. We need to work out how the structures of power and privilege in the present can be revolutionized to create a sustainable, flourishing and democratic society.

Digital technology, at least as much as any innovation over the last two centuries, offers us the opportunity to create a society that can meet the needs of every human being and allow them to explore their potential. But at present, too much power over the development of technology rests in the hands of technology capitalists and political elites who do their bidding. These people are good at what they do and also at convincing us that they are the best people to do it. They talk about egalitarianism and social connection in their public relations pitches and marketing campaigns, but what they hope to gain from digital technology is different: they aspire to wealth and power.

If we are to think of alternatives to the heedless, vampire-like drive of technology capitalism, there are examples from history that are worthy of consideration. In looking back we should not dwell on the elites, who tend to dominate the social and cultural life of each generation by default rather than through any test of merit. By far the most interesting ideas about how to build alternative futures have instead come from below. Our past is full of these kinds of struggles, though they tend to be less well documented than the great-men-of-history narratives. People have been reinventing the received wisdom on how the world works for generations, and there is much to learn from these movements and struggles.

One iconic example comes from the tumultuous nineteenth century, just a few years before Bellamy published *Looking Backward*: the Paris Commune of 1871. It was a time when the French Revolution still loomed large in people's memories, and industrialization was wreaking havoc on traditional ways of doing things—not unlike how digital technology is transforming industrial production and social relationships today. As Bertell Ollman put it, "some were delighted by these developments, most were appalled, and everyone was amazed." These transformations destabilized the political balance of the continent. By 1871, the people of Paris were thoroughly sick of bearing the costs of imperial wars,which disproportionately fell to the poor and working people of France. Their new president, Adolphe Thiers, sensed the importance of imposing centralized control over the city, as it became clear that activists were gaining political momentum in the nation's capital by putting forth a more radical and democratic vision of a republic.

Paris became the epicenter of a face-off between the old ruling regime and the discontented workers. After suffering through a particularly bitter four-month-long siege at the hands of German armies, the local National Guard sensed the moment of disquiet and began to organize themselves into an alternative form of government. Ensconced among the radical residents of Paris, the National Guard refused orders from Thiers to disarm the people. The situation became so unstable that Thiers decided to abandon Paris and set up government in Versailles, and the entire elite fled the city to follow him. The Communards, as they called themselves, took power.

In the months that followed, they turned Paris into an autonomously organized society, a commune that experimented with alternative ways of structuring social and political life based on collaboration and cooperation.

These improvised forms of self-government were so bold and transformative that they still provoke discussion and debate, a century and half later. They provide invaluable raw material for understanding alternatives to a society built on economic class and civil representation, opening up possibilities for deeper engagement with the ideas of democracy and equality that are worth examining today.

Who were the Communards exactly? Surveys taken later found that only a small percentage of them were professionals or small business owners. A full 84 percent of those left in control of Paris were manual laborers or wage earners. In this context, what happened next was probably a surprise to Thiers and his fellow elite, who showed contempt for the rabble they considered below them in social status. The Commune government addressed their constituents: "You are masters of your destiny. Strong in your support, the representatives that you have just established will repair the disasters caused by the fallen powers." Their manifesto was plastered on the walls of Paris:

> The recognition and consolidation of the Republic, the only form of Government compatible with the rights of the people and the regular and free development of society ... The absolute guarantee of individual liberty, of liberty of conscience and liberty of labour... It is the end of the old governmental and clerical world, of militarism, of monopolies, of privileges to which the proletariat owes its servitude, the country its miseries and disasters.

True to this promise, the Commune unleashed some of the most radical ideas of participatory democracy the world had ever seen.

The Commune abolished conscription and the standing army; it decreed the separation of church and state, and converted all church property into national property; it declared education free, fixed the salaries of all judicial and associated justice workers, liberated political prisoners, suspended the payment of all rents for six months, publicly burned the guillotine, closed pawn shops, and abolished night work for bakers. That was just the beginning.

Workers began to organize their premises in ways that were decentralized and locally governed, with democratically elected administrators setting wages and enforcing limits on the length of the working day. They demanded equality and autonomy: "fair remuneration for their work, freedom of public assembly and the election of their own Municipal Council." The women of Paris embraced the opportunity for change presented by the Commune. They formed a Union des Femmes, advocating for equal pay for women working in factories in Paris, and were given significant public responsibilities by the Commune. It was the Commune's largest and most effective organization. Women achieved positions of power in the administration of the Commune that were unprecedented for the period. Marriage laws were liberalized, the administration offered financial support for both legitimate and illegitimate children of soldiers, and women even adapted their attire to suit the revolutionary moment. A Federation of Artists was formed, democratizing the administration of the fine arts in Paris and advocating for the funding of promising young artists who challenged establishment tastes. The president of the federation, the painter Gustave Courbet, summarized the prevailing feeling: "Paris is a true paradise … All social groups have established themselves as federations and are masters of their own fate."

The central motivation of the Commune was the urge to improve the lives of the poor. These people had lived side by side before they took control of the city. They had fought as soldiers; they had worked together in factories; they had experienced the daily humiliation of being poor in Parisian society, beholden to the whims of the elite. Through the process of struggle, they found they had a common notion of the enemy but they also began to develop shared ideas of how their skills and ingenuity could be put to use.

This outcome was hardly planned or expected. In writing about the Paris Commune, Marx talks about how the working class "did not expect miracles." They did not have a plan or destination in mind when they tore down the oppressive society they found themselves in. They had "no ready-made utopias to introduce by decree of the people," Marx wrote. But the Communards were a living example of how "any conditions which arise in historical time are capable of disappearing in historical time." They showed that it was possible for ordinary people to seize control of their own destiny, without the need for technocratic, religious or wealthy elites presiding over social affairs. As professor Kristin Ross puts it: "The Communards had not decreed or proclaimed the abolishment of the state. Rather, they had set about, step by step, dismantling, in the short time they had, all of its bureaucratic underpinnings." The outcome was visionary for the time, Ross says, but (I would argue) also by today's standards: "it produced a greater philosophy of freedom than either the Declaration of Independence or the Declaration of the Rights of Man, because it was *concrete*." The Paris Commune showed how cities and communities can be rebuilt in radical new ways, not by "innovating" their way out of social problems, but by empowering people to make decisions collectively.

Power in the hands of those disenfranchised by the system of industrial capitalism opened up a whole new way of doing things, where everything from work to social life and even artistic expression could be arranged differently. The process of struggle, Marx wrote, "set free the elements of the new society with which old collapsing bourgeois society itself is pregnant." In the rubble of the old structures from which the Communards were trying to break free, in the daily life of sharing, solidarity and attempts to build a new society, they found a map to a new way of existing that until then they had only conceived of in their imaginations.

Ultimately the Commune was crushed. The "Bloody Week" in which Thiers, with a strengthened army, reconquered Paris resulted in around 25,000 Parisians and 877 soldiers slain. Thousands more Communards were imprisoned or deported to New Caledonia. As Carolyn Eichner put it: "The virulence of Versailles's violent repression of the Commune reflected the enormity of its perceived threat to existing gender, class, and religious hierarchies." It would be remiss to downplay such tragedies, and wrong to pretend that everything the Communards did was right or noble. But it would be foolish to equate the failings of the Communards with the catastrophic repression by Thiers. It would be equally unwise to judge the Commune's historical value in the light of its defeat.

The Commune tells us that it is possible to think of other ways of organizing, that alternative practices and principles for running society can spring forth, quite quickly, from the existing fabric of relationships and communities. Ordinary people can cooperate to render previously unshakable convictions about how things get done as fragile as gossamer. Working people are not stupid or gormless, and they require no "visionary" technologists to tell them how to run society. They have ideas of their own, and often they are pretty good ones.

Compare the ideas that emerged from the Paris Commune to, say, the thoughts of Marc Andreesson, the venture capitalist and cofounder of Mosaic and Netscape. Andreesson has long been idolized as a genius of Silicon Valley. "No one has done more than Marc Andreessen to change the way we communicate," writes Chris Anderson in his modestly titled profile, "The Man Who Makes the Future." Andreesson has been described as "an evangelist for the church of technology." He is not a utopian in the formal sense and does not describe himself as such, but he does believe that advancements in technology in their current form are beneficial, and he has done some thinking about what this might look like in the future.

In 2014, Andreesson wrote about his vision for the world of tomorrow using a thought experiment. "Imagine a world in which all material needs are provided for free, by robots and material synthesizers." This kind of reality could solve some of the world's largest problems, a consumer utopia where people spend time pursuing creative, cultural and scientific ambitions, Andreesson claimed. At base, however, his vision was little more than a rather drab idea of ultra-capitalism: "I am not talking about Marxism or communism, I'm talking about democratic capitalism to the nth degree. Nor am I postulating the end of money or competition or status seeking or will to power, rather the full extrapolation of each of those."

Andreesson sees many of our collective problems, including inequality and unemployment, as functions of a society that clings to old ways of regulating and incentivizing economic activity. For him, a more fulsome embrace of technology capitalism will serve to strengthen society. Rather like the first wave of technological utopians, the folks who love creative destruction are also remarkably conservative in their thinking. "Firms trumpet their boldness, yet they often follow one another, lemming-like, pursuing the latest innovation," writes Tad Friend, about the venture capitalist firms of Silicon Valley, including the one that Andreesson founded. A tendency toward utopian thinking about technology makes it logical for someone like Andreesson to argue for more capitalism to solve the problems of capitalism.

In other words, utopian thinking can simultaneously present both a fantastically transcendent vision of tomorrow and a totally paralyzed understanding of the present, which obliterate the possibility of a genuinely different future. Bertell Ollman explains:

> Utopian thinking presents us with consequences (the ideal) without causes, i.e., causes capable of producing such consequences—and therefore too with causes (what exists now) that have no apparent consequences. It is not a matter of the present losing some of its potential; its entire future dimension has been wiped out. Hence, it is not only the future that gets distorted in utopian thinking but also the present. It is futureless because it does not itself exist as a cause of its own future.

Utopianism erases our understanding of the present in a vision of the future that is supposedly detached from, yet ends up hopelessly bound to, the problems of the present. It is for this reason that some of the people lauded as the most visionary in our society end up having some of the most mundane ideas. Technology capitalists love to talk up the sparkling possibilities of technology—of unleashing potential in an interconnected society. But often what is revealed in these manifestos is little more than an unambitious extension of the status quo.

On one level, this is unsurprising: those who are satisfied with the basics of the present, who might quibble with the pace of change rather than its direction, often come up against the limits of their imagination. Their technological utopias either fail to come to terms with the root source of social ills or descend into sheer fantasy. This analysis underpinned the conclusion drawn by the British artist and socialist William Morris in his 1889 review of Bellamy's *Looking Backward*. "The only safe way of reading a utopia is to consider it as the expression of the temperament of its author," Morris observed. He found the book to be, like its author, "unhistoric and unartistic" and "perfectly satisfied with modern civilisation." Morris is today perhaps best known for his sumptuous, neomedieval decorative designs, but he was also committed to the struggles of the working class throughout his life, using his status and success to support and advocate for these movements from below. Through this engagement he saw the possibilities of a new kind of society and imagined the present changing the course of the future through political struggle. According to Morris, rather than being a representation of self-government or a dismantling of class, Bellamy's utopia was a state-managed monopoly economy, managed by "the industrious, *professional* middle-class." In short, Morris concluded: "Mr. Bellamy's ideas of life are curiously limited."

Morris was a critic worth listening to. He was, as noted by Kristin Ross, one of the "foremost British supporters of the memory of the Paris Commune." Having seen the scope of possibilities unleashed in Paris, Morris spent many years helping to define the Commune's legacy, which he saw as a central question of his day. While he was pleased that Bellamy's work had inspired so many people to take an interest in the idea of socialism, he objected to its possibilities being framed as narrowly as they were in *Looking Backward*. In contrast to Bellamy's self-satisfied, unimaginative vision, Morris affirmed that

> there are some Socialists who ... [believe] it will be necessary for the unit of administration to be small enough for every citizen to feel himself responsible for its details, and be interested in them; that individual men cannot shuffle off the business of life on to the shoulders of an abstraction called the State, but must deal with it in conscious association with each other. That variety of life is as much an aim of true Communism as equality of condition, and that nothing but a union of these two will bring about real freedom.

Only by structuring society in ways that give everyday people genuine economic and social control over their lives, without subordination to some professional class of technocrats or capitalists, can we give rise to real freedom. For Morris, the great lesson of the Paris Commune was the possibility of this kind of society.

Capitalists and those content with our current form of civilization may struggle to imagine how society can meaningfully overthrow injustice, misery and waste. But it *is* possible for others to do this. It is possible for the present to be reclaimed as a cause of its own future, as Ollman might put it, if we think through how a future unlike the one promised by technology capitalism might come about.

The entire Internet and plenty of the basics of modern computing are based on free access to the infrastructure of digital technology. They are underpinned by a rejection of the idea of singular ownership and by a sense of the importance of collectively determining the basic standards that allow people to use, work and experiment together. For example, the World Wide Web Consortium (W3C) is an international community of organizations and staff that work with the public to set the standards of the web. Its mission is "to lead the Web to its full potential" by defining standards around web technologies "designed to promote consensus, fairness, public accountability, and quality." The W3C is by no means perfect and has its own schisms and conflicts. But it is organic, in the sense that it is a form of governance designed to be collaborative and interactive, which incidentally reflects how many people work within the technology industry already. In other words, the developments that gave rise to technology capitalism also allow for the possibility of structuring digital life in alternative ways. As Matt Bruenig observes:

> Were all of these tools developed along the same proprietary lines as the software that comes out of Microsoft or Apple, the web (and the ability to innovate on it) would look very different. You couldn't just start a Facebook in a dorm room; you'd need to shell out big money to pay for all the bits of software that go into a web server stack. The nearly frictionless ability to, if you know what you are doing, push out new internet-based tech products is solely the provenance of a community of sharing-driven programmers ideologically opposed, in varying degrees, to the capitalist ethic.

Something communal and collective has been built right alongside the technology that has developed in the individualistic, atomized pursuit of profit. So while we have Internet behemoths like Facebook that monetize our social interactions, we also have its raggedy equivalent in the form of craigslist, one of the most popular sites in the world, once described as "gigantic in scale and totally resistant to business cooperation." As Google seems to become ever more omnipotent, tiny competitors in the search engine market like DuckDuckGo attract users by pledging not to track them, collect their data or filter results into a personalized bubble. Just like we have major telecommunication companies cooperating with government to facilitate mass surveillance, we also have nonprofit software groups making free and open source applications to protect citizen communications with encryption. The market for web platform design and content management has produced enormous and profitable companies, but it has also produced WordPress and

Drupal, open source services for blogging and content management respectively. These services are used by nonprofits and community organizations all over the globe and, as Rebecca Mackinnon points out, "are developed, maintained, and upgraded by a community of volunteer developers."

Right alongside these efforts are the workers within technology capitalism doing their best to beat out spaces of resistance in the industry itself. Waves of radicalism are now eroding business as usual in Silicon Valley, as workers draw connections between the technology they build and the worst excesses of government and corporate power. Technology workers are creating a grassroots movement centered on a refusal to build technologies of violence and oppression, in favor of a world powered by technologies of peace and justice. Organizations such as the Tech Workers Coalition rally around workplace issues, as well as protesting outside Peter Thiel's company, Palantir. Similar groups are springing up in India and Brazil. They organize around industrial issues and broader political ones and actively draw the connection between the two.

These activists do not toil in the factories of Paris, nor do they build barricades out of cobblestones to keep the armies of the elite out of the communities they have created. But the people who make technology and the many more who use technology do often have a better grasp of what kind of alternative futures are possible if they are given the chance to put it into practice. What technology capitalists therefore produce, above all, are their own gravediggers.

6

Collaborative Work Is Liberating and Effective

Poetical Philosophy, from Lovelace to Linux

From the moment she was born, Augusta Ada Gordon was discouraged from writing poetry. It was a struggle against her genetic predisposition. Her father had led by example in the worst possible way, cavorting around the Mediterranean, leaving whispered tales of deviant eroticism and madness wherever he went. He penned epic stanzas full of thundering drama and licentiousness. Lord Byron understood the dangers of poetry. "Above all, I hope she is not *poetical*," he declared upon his daughter's birth; "the price paid for such advantages, if advantages they be, is such as to make me pray that my child may escape them." Ada, as she was known, failed to make this escape and barely enjoyed the advantages. The poetry she went on to write was beyond even her father's imaginings.

Likewise hoping that her daughter might avoid the fate of a father who was "mad, bad and dangerous to know," Ada's mother, Lady Anne Isabella Milbanke, ensured that she was schooled with precision and discipline in mathematics from her earliest days, and closely watched her for any signs of the troubles that had plagued her father. Lord Byron, abandoning them weeks after Ada's birth, died when she was eight; his legacy cast a ghostly shadow over her life.

Ada's schooling marched ever forward, toward an understanding of the world based on numbers. "We desire certainty not uncertainty, science not art," she was insistently told by one of her tutors, William Frend. Another tutor was the mathematician and logician Augustus De Morgan, who cautioned Ada's mother on the perils of teaching mathematics to women: "All women who have published mathematics hitherto have shown knowledge, and the power of getting, but ... [none] has wrestled with difficulties and shown a man's strength in getting over them," he wrote. "The reason is obvious: the very great tension of mind which they require is beyond the strength of a woman's physical power of application." But Ada was never going to be denied the opportunity to learn about mathematics. Lady Anne was a talent herself, dubbed "Princess of Parallelograms" by Lord Byron. Having managed to outlive him, the desire to expunge in her daughter the slightest genetic tendency for mad genius and kinky sex took precedence over any concerns about Ada's feminine delicacy.

Married at nineteen years of age, Ada, now Countess Lovelace, demonstrated curiosity and agility of mind that would prove to be of great service to the world. Just a year before her marriage, in 1833, she had met Charles Babbage, a notable mathematician with a crankish disposition (he could not stand music, apparently, and started a campaign against street musicians). Together they worked on plans for the Analytical Engine, the world's first mechanical computer. It was designed to be a mechanical calculator, with punch cards for inputting data and a printer for transcribing solutions to a range of mathematical functions. Babbage was a grand intellect, with a penchant for snobbery and indifference to many of the practicalities of getting things done. Lovelace was his intellectual equal but arguably better adapted to social life.

Like the proverbial genius, Babbage struggled with deadlines and formalities. When one of his speeches was transcribed for publication in Italian and neglected by Babbage, Lovelace picked it up and translated it. She redrafted parts of it to provide explanations to the reader. Her work ended up accounting for about two-thirds of the total text. This became her significant contribution to the advancement of computing: turning the transcription into the first-ever paper on computer science. It became a treatise on the work she and Babbage did together.

There remains some controversy about the extent of Lovelace's participation in this project, but ample historical evidence exists to dismiss the detractors, not least the direct praise bestowed on her work and intellect by Babbage. Lovelace applied her mathematical imagination to the plans for the Analytical Engine and Babbage's vision of its potential. She sketched out the possibility of using the machine to perform all sorts of tasks beyond number crunching. In her inspired graphic history of Babbage and Lovelace, Sydney Padua describes Lovelace's original contribution as one that is foundational to the field of computer science: "By manipulating symbols according to rules, *any* kind of information, not only numbers, can be operated on by automatic processes." Lovelace had made the leap from calculation to computation.

Padua describes the relationship between Babbage and Lovelace as complementary in computational terms. "The stubborn, rigid Babbage and mercurial, airy Lovelace embody the division between hardware and software." Babbage built the mechanics and tinkered endlessly with the physical design; Lovelace was more interested in manipulating the machine's basic functions using algorithmic formulas. They were, in essence, the first computer geeks.

The kind of thinking needed to build computers is precisely this combination of artistry and engineering, of practical mechanics and abstract mathematics, coupled with an endless curiosity and desire for improvement. The pioneering pair's work blurred the division between science and art and navigated the spectrum between certainty and uncertainty. Without Babbage, none of it would have happened. But with Lovelace's predilection for imaginative thinking and education in mathematics, a perfect alignment of intellect allowed for the creation of computer science. Lovelace and Babbage's achievements were impressive because they challenged what was possible while at the same time remaining grounded in human knowledge.

And beyond all this, Lovelace was a woman. (A woman!) In direct contradiction to her tutors' warnings decades earlier, Lovelace,

Babbage wrote, was an "enchantress who has thrown her magical spell around the most abstract of Sciences and has grasped it with a force which few masculine intellects (in our country at least) could have exerted over it." Lovelace showed it was possible to transcend not only the bounds of orthodox mathematics but also her socially prescribed gender role.

No doubt all this caused Lovelace's mother considerable worry. The madness seemed to be catching up, much to her consternation. In the years after her visionary publication, Lovelace poignantly beseeched Lady Anne: "You will not concede me philosophical poetry. Invert the order! Will you give me *poetical philosophy, poetical science*?"

For Babbage, perfect was the enemy of good, and he never did manage to build a full model of his designs. In 1843, knowing that he struggled with such matters, Lovelace offered, in a lengthy and thoughtful letter, to take over management of the practical and public aspects of his work. He rejected her overtures outright yet seemed incapable of doing himself what was required to bring his ideas to fruition.

Lovelace's work in dispelling myths and transforming philosophy was cut short when she died of cancer aged just thirty-six. Babbage died, a bitter and disappointed old man, just shy of eighty. The first computers were not built until a century later.

Technological advances are a product of social context as much as of an individual inventor. The extent to which innovations are possible will depend on a number of factors external to the individuals who make them, including the education available to them, the resources they have to explore their ideas, and the cultural tolerance for the kind of experimentation necessary to develop those ideas. Melvin Kranzberg, the great historian of technology, observed that "technology is a very human activity—and so is the history of technology." Humans are responsible for technological development but do not labor in conditions of their own choosing. Had Babbage been a bit more of a practical person, in social as well as technological matters, the world may not have needed to wait an extra century for his ideas to catch on. Had Lovelace lived in a time where women's involvement in science and technology was encouraged, she might have advanced the field of computer science to a considerably greater degree.

So too, then, technological developments more generally can only really be understood by looking at the historical context in which they occur. The industrial revolution saw great advances in production, for example, allowing an economic output that would scarcely be thought possible in the agrarian society that had prevailed a few generations earlier. These breakthroughs in technology, from the loom to the steam engine, seemed to herald a new age of humanity in which dominance over nature was within reach. The reliance on mysticism and the idea that spiritual devotion would be rewarded with human advancement were losing relevance. The development of technology transformed humanity's relationship with the natural world, a process that escalated dramatically in the nineteenth century. Humans created a world where we could increasingly determine our own destiny.

But such advances were also a method by which workers were robbed of their agency and relegated to meaningless, repetitive labor without craftsmanship. As machines were built to do work traditionally done by humans, humans themselves started to feel more like machines. It is not difficult to empathize with the Luddites in the early nineteenth century, smashing the machines that had reduced their labor to automated work. In resisting technological progress, workers were resisting the separation of their work from themselves. This separation stripped them of what they understood to be their human essence. For, whatever the horrors of feudalism, it allowed those who labored to see what they themselves produced, to understand their value in terms of output directly. Such work was defined, at least to a certain extent, by the human creativity and commitment around it. With industrialization and the atomization of craftsmanship, all this began to evaporate, absorbed into steam and fused into steel. Human bodies became a vehicle for energy transfer, a mere input into the machinery of production. It gave poetic significance to the term Karl Marx coined for capital: dead labor.

Though the Luddites are often only glibly referenced in modern debates, the truth is that they were directly concerned with conditions of labor, rather than mindless machine-breaking or some reactionary desire to turn back time. They sought to redefine their relationship with technology in a way that resisted dehumanization. "Luddites opposed the use of machines whose purpose was to reduce production costs," writes historian Kevin Binfield, "whether the cost reductions were achieved by decreasing wages or the number of hours worked." They objected to machinery that made poor-quality products, and they wanted workers to be properly trained and paid. Their chosen tactic was industrial sabotage, and when their frame-breaking became the focus of proposed criminal law reform, it was, of all people, Lord Byron who leaped to their defense in his maiden speech to the House of Lords. Byron pleaded that these instances of violence "have arisen from circumstances of the most unparalleled distress." "Nothing but absolute want," he fulminated, "could have driven a large and once honest and industrious body of the people into the commission of excesses so hazardous to themselves, their families, and the community."

The historical effect of this strategy has been to associate Luddites forever with nostalgia and a doomed wish to unwind the advances of humanity. But to see them as backward-looking would be an interpretive mistake. In their writings, the Luddites appear more like a nineteenth-century equivalent of Anonymous: "The Remedy for you is Shor Destruction Without Detection," they wrote in a letter to the home secretary in 1812. "Prepaire for thy Departure and Recommend the same to thy friends."

There is something very modern about the Luddites. They serve as a reminder of how many of our current dilemmas about technology raise themes that have consistently cropped up throughout history. One of Kranzberg's six laws of technology is that technology is neither inherently good nor bad, nor is it neutral. How technology is developed and in whose interests it is deployed is a function of politics. The call to arms of the Luddites can be heard a full two centuries later, demanding that we think carefully about the relationship between technology and labor. Is it possible to resist technological advancement without becoming regressive? How can the advances of technology be directed to the service of humanity? Is work an expression of our human essence or a measure of our productivity—and can it be both?

Central to understanding these conundrums is the idea of alienation. Humans, through their labor, materially transform the surrounding world. The capacity to labor beyond the bare necessities for survival gives work a distinct and profound meaning for human beings. "Man produces himself not only intellectually, in his consciousness, but actively and actually," Marx wrote, "and he can therefore contemplate himself in a world he himself has created." Our impact on the world can be seen in the product of our labor, a deeply personal experience. How this is organized in society has consequences for our understanding of our own humanity.

What happens to this excess of production—or surplus value—is one of the ultimate political and moral questions facing humanity. Marx's critique of capitalism was in essence that this surplus value unfairly flows to the owners of capital or bourgeoisie, not to the workers who actually produce it. The owning class deserve no such privilege; their rapacious, insatiable quest for profit has turned them into monstrous rulers. Production becomes entirely oriented to their need for power and luxury, rather than the needs of human society.

Unsurprisingly, Marx reserved some of his sharpest polemical passages for the bourgeoisie. In his view, the bourgeoisie "resolved personal worth into exchange value, and in place of the numberless indefeasible chartered freedoms, has set up that single, unconscionable freedom—Free Trade. In one word, for exploitation, veiled by religious and political illusions, it has substituted naked, shameless, direct, brutal exploitation."

This experience of exploitation gives rise to a separation or distancing of the worker from the product of her labor. Labor power becomes something to be sold in the market for sustenance, confined to dull and repetitive tasks, distant from an authentic sense of self. It renders a human being as little more than an input, a cog, a calculable resource in the machinery of production. For those observing the development of the industrial revolution, this sense of alienation is often bound up with Marx's analysis of technology. The development of technology facilitated the separation between human essence in the form of productive labor and the outputs of that labor. Instead workers received a wage, a crass substitute for their blood, sweat and tears, a cheap exchange for craftsmanship and care. Wages represented the commodification of time—they were payment for the ingenuity put into work. The transactional nature of this relationship had consequences. "In tearing away from man the object of his production," Marx wrote, "estranged labor tears from him his *species-life*, his real objectivity as a member of the species, and transforms his advantage over animals into the disadvantage that his inorganic body, nature, is taken from him."

As Amy Wendling notes, it is unsurprising that Marx studied science. He sought to understand the world as it is, rather than pursue enlightenment in the form of spirituality or philosophy alone. He understood capitalism as unleashing misery on the working class in a way that was reprehensible but also as Wendling put it, "a step, if treacherous, towards liberation." There was no going back to an agrarian society that valued artisan labor. Nor should there be; in some specific ways, the industrial revolution represented a form of productive progress. But how things were then were not how they could or should be forever. Marx's thinking was a product of a desire to learn about the world in material terms while maintaining a vision of how this experience could be transcended. Navigating how to go forward in a way that valued fairness and dignity became a pressing concern of many working people and political radicals in his time, a tradition that continues today.

Tensions between technology and labor—the objective and subjective, the creative and the analytical—are all around us under capitalism. In the digital age, one field is particularly relevant: the history of software development. In it, we can see how labor can be both unalienated and productive, and how it can be limited by the imposition of a profit motive. In a world in which digital technology has great potential for helping us organize human activity efficiently and sustainably, the limits placed on it by capitalist modes of production are worth examining.

For much of the twentieth century, computer programming was not the vastly profitable industry that it is today. (Babbage and Lovelace would point out that it was even less profitable in the nineteenth.) The modern industry of computing started out as niche projects in universities and large industrial companies, or as experimental projects by eccentric individuals. Geeks have always loved pet projects. No doubt the parents of Konrad Zuse were thrilled when he built the first modern mechanical computer in their living room in Berlin in 1936. Zuse was known for being "obsessed" with his machines. He labored in conditions described as "almost total intellectual isolation" (it is unclear if this particular biographer is impressed, sympathetic or both). Apolitical by nature, he appeared to suffer World War Two as something of an irritating distraction from his endless tinkering. As Lovelace observed all too well in Babbage, geekery to some extent has always involved a struggle between the antisocial personality traits that tend to drive the pursuit of such projects and the limitations on zeal imposed by the reality of modern society.

By 1953, IBM had built the first electric computer that was mass-produced. In these early days of computing, up until the late 1970s, the relationship between hardware and software was quite different to how we understand it today. Hardware manufacturers would give away software for free. IBM was one such example: it supplied software as a standard extra when it sold its hardware. The software was "free," meaning without charge, but also in the sense that IBM encouraged experimentation with its products. Users would provide feedback to IBM to help it make changes and fix problems in the system. If anything, users were streets ahead of IBM in improving the software, the hive minds of geeky college students constantly frustrated at the plodding pace of the batch-processing-obsessed developers at IBM.

The software IBM distributed was nonetheless subject to formal copyright protection. It was considered a work of authorship, which imposed some limitations on how it could be used. But this was really only a theoretical matter. In practice, professor Eben Moglen describes, "Mainframe software was cooperatively developed by the dominant hardware manufacturer and its technically sophisticated users, employing the manufacturer's distribution resources to propagate the resulting improvements through the user community." In other words, it was a world of collaborative innovation, a constant feedback loop. This sits in contrast to modern understandings of property, which see software as a discrete thing to be owned under copyright. "The right to exclude others, one of the most important 'sticks in the bundle' of property rights," Moglen notes, "was practically unimportant, or even undesirable."

Stephen Levy, in his overly affectionate if thoughtful portrait of this world, *Hackers*, published in 1986, offers an insight into the nature of collective innovation in the student computer labs at places like MIT. Students would write all sorts of programs and distribute them widely without expecting anything in return. In fact, they would invite collaboration—leaving the first iteration of a program in a communal storage drawer for others to pick up and work on. One computer company ended up using a program written by MIT students—probably the first ever computer game, called Spacewar—as a diagnostic testing tool before shipment, leaving the game on there for unsuspecting customers to stumble across. Spacewar, like all other programs, was a collaborative project; the framework of the code was written by one person, then added to by others, until eventually students had rigged up the computer lab lighting to synchronize with key moments in the game. Here was creative design undertaken collectively.

It was clear to computer manufacturers that collaborative work on software like this resulted in a superior product for users. The community of programmers worked together to address problems and needs they had as individuals, writing code to create solutions that were fed back to the manufacturer. The distinction between user and programmer was blurred, even porous. This kind of voluntary collective work was (and to a certain extent still is) known as hacking. Levy describes hacking as pulling things apart, changing parts and seeing how these changes affected the system overall, with the ultimate aim of putting the various parts to optimal use. Hackers possessed a "kind of restless curiosity," he wrote; their projects were undertaken not necessarily for profitable or constructive goals but for the "wild pleasure taken in mere involvement." To qualify as hacking, such work would have to be stylish and display technical innovation. The community of users involved the same people who were creating fixes for their problems, resulting in a virtuous cycle of collaborative improvement.

Such a world was not open to everyone. You had to have time, skills and devotion to the craft to succeed. This kind of meritocratic thinking allowed some people to participate in the lab in a way that broke traditional social rules. David Silver, for example, started working in the MIT lab before he had finished school. He began hanging around at just fourteen years of age, a fact many hackers probably failed to notice. What they did notice, however, was his skill with robotics, especially when he managed to build what he called the "Silver Arm"—the first robotic arm controlled by the lab's computer.

But alongside the capacity of hacker culture to discard social mores that might have prevented a kid like Silver being allowed past the front door of the lab, it cultivated a regrettable indifference to those most held back by such social strictures. Women were the most obvious example. There were very few of them active in the MIT lab. "The sad fact was that there was never a star-quality female hacker. No one knows why," wrote Levy, with a tin ear.

> There were women programmers and some of them were good, but none seemed to take to hacking … Even the substantial cultural bias against women getting into serious computing does not explain the utter lack of

> female hackers. "Cultural things are strong, but not *that* strong," [Bill] Gosper would late conclude, attributing the phenomenon to genetic, or "hardware," differences.

This absence of women was partly a function of women being underrepresented in disciplines associated with computing, since such fields were traditionally considered male preserves, including science, engineering and mathematics. But this is hardly the whole story: the reality is that many women, including Lovelace but also countless others, have made significant contributions to the history of computing.

Grace Hopper was instrumental to the field of computer programming, cutting her teeth at Harvard during World War Two and later joining other female colleagues at Pennsylvania State University. Hopper was committed to making computer programming available to non-experts. She worked on pioneering developments such as the use of compilers and the creation of the language COBOL, which remain influential today. Many African American women too did groundbreaking work at NASA, including Katherine Johnson, Dorothy Vaughan and Mary Jackson, whose experiences have been documented in the book (and film) *Hidden Figures*. These women were known as human computers, tackling enormously complex mathematical tasks and getting very little credit because of their gender and race. Together with colleagues like Margaret Hamilton, responsible for the Apollo on-board flight software around the same time, they performed work that was essential to NASA and the field of computer science more generally.

As the industry began to professionalize—formalizing its status as more elite, more specialized and deserving of more respect—women were pushed out by various gatekeepers. The idea expressed by Levy or Gosper that women were somehow biologically unsuited to this kind of work was therefore not only historically inaccurate, it also demonstrated a lack of understanding among the early hackers of how social expectations function in specific and sophisticated ways. The absence of women and the downplaying of their achievements in the field were taken for granted, due to a naive presumption that it was down to something innate or genetic, rather than to external social or political factors that could be changed.

And these prejudices have proven resilient, as shown by the popularity of the "anti-diversity" memo circulated among Google workers in 2017. This memo argued that the gender imbalance in the industry was less a function of unconscious bias than of differences in the distribution of traits between men and women. The hacker and tech communities are not immune to the structural factors that shape the world around them. Hackers have often valued the idea that they are working in a meritocracy, where worth is measured according to objective factors like skill, creativity and effort. But rather than elevating fairness, this has flattened difference. Devotion to meritocratic thinking has made it too easy to ignore how unfairness has been built in to our social systems and structures. Meritocratic thinking claims to be gender-blind, but this has effectively meant being blind to gender inequality.

These are legitimate critiques of the free-spirited culture of collaboration of the MIT lab, and they unfortunately remain relevant today. But without dismissing such concerns, it is still possible to observe in this history the raw material of alternative ways of working and organizing productive labor.

The MIT lab was a hotbed of productive activity, but it was also more than mere number crunching or code writing. Modern hacking, just like the work of Babbage and Lovelace, is at heart a creative pursuit. Levy observed this in transcribing the so-called hacker ethic he observed during the period, which included the declaration: "You can create art and beauty on a computer." It is tempting to think of computer programming as a left-brain activity, for mathematicians and logicians. But hacking is without question an artistic endeavor. It is based on aesthetics: hacking is the art of finding the most elegant solution to a particular problem.

Donald Knuth figures as something of an elder statesman of computing. First published in 1968, his foundational work is called, tellingly, *The Art of Computer Programming*. Knuth talks about computer programming as both a science and an art. He defines science as "knowledge which we understand so well that we can teach it to a computer." When we reach the limits of this knowledge, Knuth argues, we turn to art to address the mystery. He writes:

> When I speak about computer programming as an art, I am thinking primarily of it as an art *form*, in an aesthetic sense. The chief goal of my work as educator and author is to help people learn how to write *beautiful programs* ... when we prepare a program, it can be like composing poetry or music; ... programming can give us both intellectual and emotional satisfaction, because it is a real achievement to master complexity and to establish a system of consistent rules.

Hacking is work that is as much about means as it is about ends, and writing computer programs is an exercise in both art and science; it is both functional and aesthetic labor. Such labor provides the opportunity to contemplate ourselves in a world we have created, what Marx called our *species-life*.

Innovation is not epitomized by some tortured genius working alone or a billionaire who once came up with a clever idea. Some of our most radical new technological developments were a result of teamwork, drawing on multiple people's varied skill sets. Working in this way is both empowering, as we understand how we contribute to a broader project, but also effective, because this kind of collaboration between diverse minds allows us "to master complexity and to establish a system of consistent rules." Computing—one of the greatest technological advances in recent centuries—began as a small pocket of sophisticated craft labor practiced in a relatively unalienated manner, while the world of capitalist enterprise carried on all around.

But with the rise of personal computing, many programmers recognized that software could also be a source of profit. Hardware became cheaper and available in the consumer marketplace as the personal computer industry began to grow. IBM "unbundled" its products, out of antitrust concerns, selling hardware separately from the provision of software. Other companies began to follow suit. In response to these developments, some programmers began writing software independently for sale, not simply to serve the functional utility of hardware.

With this development of an industry came a critical change in the process of creating software. Proprietary software—software sold as a product, usually independent of hardware—depended on a "closed source" model, whereby the source code was concealed or closed off from the user. Software was sold in an executable format, without disclosing its underlying structure or processes. This design was required by those selling proprietary software to limit the extent to which users could simply copy it, share it and install the program, bypassing the author and payment. It represented a fundamental transformation in the process of software design, from one that transparently served the interests of users to one that served the interests of the owners of companies.

These changes in software production came with legal and political realignments. Copyright law took on new significance. Most famously, this found expression in Bill Gates's letter to his community of fellow computer hobbyists in 1976. Frustrated with colleagues who took a more liberal attitude to software sharing, Gates exhorted the community to pay up:

> As the majority of hobbyists must be aware, most of you steal your software. Hardware must be paid for, but software is something to share. Who cares if people who worked on it get paid? Is this fair? ...
>
> What hobbyist can put three man-years into programming, finding all the bugs, documenting his product and distribute it for free? ...
>
> Most directly, the thing you do is theft.

Gates's campaign to generate respect for proprietary software was not universally welcomed among the burgeoning community of hackers. Some directly opposed it. But there was no denying that Gates had a very big impact on the world of software design.

However, Gates represented only one part of the history of software. As a counterpoint to people like him, other hackers started to think more broadly about the idea of freedom in the context of their work and tried to put this into practice. One of these was Richard Stallman. He began working as a programmer at Harvard in the early 1970s and spent a happy decade of creative software development there with fellow geeks, learning and tinkering. But it was not to last. Stallman was devastated when his community fell apart as programmers were lured into private industry. Fellow developers sold programs to companies, bringing them under the strictures of copyright, making them no longer available for modification or collaborative rewriting.

This experience motivated him to create an entire ecosystem of software that was free of these influences and constraints. He called this system GNU, a word play that defined his program differently to Unix, the other main operating system at the time (GNU stood for: Gnu Not Unix).

The kind of "free" Stallman was thinking about was not a reference to the price of software, though free software was often accessible for little more than the cost of copying and posting the disks. (Today most free software is downloadable without charge.) Rather, people like Stallman were concerned with freedom in the developmental environment. Stallman describes it as free as in "free speech," not as in "free beer." "Software sellers want to divide the users and conquer them, making each user agree not to share with others," Stallman wrote in his manifesto, first published in 1983. "I refuse to break solidarity with other users in this way." Stallman called for volunteers to assist in the project, and many joined him.

The kind of freedom these hackers were after concerned the conditions of labor and the liberty to collaborate. It was about the promise of creative work without the necessity of working for wages. Again, such collaboration was not possible for everyone, as participants had to have the economic means and be taught certain skills to be involved. But it was, at its core, the beginning of a movement that aimed to dismantle the alienation associated with proprietary software production and use.

In resisting the idea of closed, proprietary software by means of open, liberated software, these hackers shared a common objective with the Luddites over two centuries earlier. The Luddites struggled against the attacks on the integrity of their work that were occasioned by industrialized production techniques. The hackers similarly refused the prospect of being made to perform shoddy work in closed corporate environments, and they devoted themselves to undermining the power of the proprietary software system. Luddites broke frames; hackers wrote open source code and continue to do so today.

One of the most interesting strategies in this movement was its use of the law. Stallman worked with lawyers to devise legal protection for free software development. He published the code for GNU programs, still as a species of copyright but under a form of license known as the General Public License (GPL). The GPL is what is called a "permissive" license. Its stated aim is "to guarantee your freedom to share and change all versions of a program—to make sure it remains free software for all its users." It essentially serves to protect the openness or liberated nature of a piece of software by ensuring the code is available to all users into the future. The terms of the GPL permit sharing, modification, hacking, and even combining GNU programs with proprietary products for sale. The only requirement is that *all* of the subsequent source code be also released under the GPL. This feature of the GPL is often labeled "viral": it serves to protect free software from proprietary expropriation but also to build and expand the digital commons by ensuring that future programs using free software are also publicly available. In short, the GPL takes the traditional view of copyright and inverts it.

The object of the GPL, together with other permissive licensing arrangements inspired by its example, is to reverse the rights of the author and protect the freedom of the user. Together these licenses are referred to as "copyleft." Copyleft is often accompanied by the tag "all rights reversed," an inversion of the traditional notion of copyright as "all rights reserved." It broke down the distinction between author and user, allowing these roles to be much more fluid than could ever be possible with closed source, proprietary software.

Stallman may have been the first to think about these issues and give them legal expression. But he was not the only one. He was representative of a broader and growing movement of hackers, geeks and hobbyists, all tinkering with the same early computer programs and working together to improve them all over the world. The best example of this was the development of the most complex part of any operating system, the kernel, which acts as the facilitator between the hardware and software applications. Stallman had delayed building it: constructing a free kernel was going to be a tricky, time-consuming task of monumental proportions. It was the last and missing piece in Stallman's vision of a complete set of free software.

In 1979, one of the main kernels available at the time, Unix, changed its licensing arrangements to prohibit users from reading or modifying the source code. The decision to close the code for Unix was made by AT&T (who owned the code) when an antitrust ruling expired that had prevented them from selling the product commercially. AT&T got down to the business of selling Unix, and, as such, it was no longer an open source product available for people to tinker with and modify. With Unix out of reach for hackers, an academic named Andrew Tanenbaum, at the Free University in Amsterdam, created a mini-kernel for his students to learn from, called Minix. The source code was freely available and quickly became popular among the growing community of hackers with access to more widely available computers, looking for alternative programs to innovate, play with and improve. However, Tanenbaum was not interested in taking the many suggestions he received on how to improve Minix, as it was designed as an educational program, and he wanted it to be kept simple. The hackers among Minix users, constantly itching to test the potential of their computers, began to feel frustrated.

One such hacker, Finnish computer science student Linus Torvalds, resolved instead to just start from scratch and create a new kernel. He called it Linux. In part because of the ambitious size of the project, but also the school of hackers in which he swam, Torvalds made the source code of Linux freely available. He posted his project to the community board and invited feedback. Shortly after, Torvalds licensed it under the GPL, as Stallman had done. Linux became another foundational part of the digital commons of free software. It supplied the missing piece in Stallman's project of a complete operating system, ultimately known as GNU/Linux.

This collective creation was the product of a community devoted to sharing and collaborating. Torvalds could have taken a closed software path for Linux. He could have kept the source code to himself, working with others to refine his product and take it to market. It is easy to imagine that this would have proven highly profitable for him personally. Yet, interestingly, he described this choice in the following terms: "Making Linux freely available is the *single* best decision I've ever made." Torvalds understood the design advantages of free software, the most obvious of which is that people provide feedback on the features they want, so the product is designed for and services them. But there are other important structural features in this design process. Many minds work on finding bugs and fixing problems, meaning that improvement happens at a rapid rate. Consumers of the product use it, find bugs and report them, launching a new cycle of production. Such an environment is reminiscent of the early days at IBM and the MIT computer lab. It was a progressive adaptation of the original hacker style.

Linux, with the help of the Internet, was worked on round the clock in a multinational project of collaborative improvement. The result was that Linux developed at an astonishing pace; the sheer number of people and hours devoted to its expansion and improvement meant that it overcame design inferiorities, to a high standard. "Everybody puts in effort into making Linux better, and everybody gets everybody else's effort back," Torvalds explained. "And that's what makes Linux so good: you put in something, and that effort *multiplies*."

The outcome of the free software movement was an entire software ecosystem, produced in conditions of openness and freedom, protected by copyleft licensing. Many thousands of users, both seasoned hackers and everyday people, have contributed to creating GNU/Linux. The project involved countless hours of human work, contributed usually for free, coordinated over decades across several continents. Many thousands of other programs are also licensed under the GPL, often produced in similar circumstances. The resulting body of software is considered superior in design and performance to many commercial products, even by its competitors. Linux, for example, has gone on to become the kernel that forms the backbone of many operating systems for both supercomputers and android consumer products. While free software programs are used in many consumer products sold in the marketplace, their component parts are available for zero or very limited cost. This represents an objectively stunning achievement, not only in terms of the quality of the output but also logistically.

The hacker ethic found an adversary in the development of the proprietary software industry. But it was not displaced. It took on new life in hackers who worked on projects motivated by fun, by the sheer joy that flows from creating something that is useful. It serves as a living, breathing refutation of Bill Gates's haughty assumption that creativity and ingenuity are motivated by the prospect of material gain. It undercuts orthodox economists' understanding of the free rider problem and lays bare mistaken beliefs about the selfishness of human nature. It contradicts the idea that only experts or the highly trained can craft beautiful things, and it suggests that the concept of a single author may not be the best way to structure work that advances humanity.

And, like many movements that have come before it, the free software movement demonstrates directly the power of working as a collective and orienting the resulting products to the service of users. The diversity of contributors, much like the varied skills of Lovelace and Babbage, propels us to new heights in technological terms. The software that has been produced by the free software movement is transparent, accountable and adaptable. It also challenges the legitimacy of the enormous profits raked in annually by an entire industry.

Volkswagen probably never expected to be found out. The car manufacturer admitted in late 2015 to systematically using software to cheat environmental regulation. In testing, everything had looked fine. But once on the road, their cars performed better—and the emissions levels were well above legal limits. The software in the car, it seems, was designed to pick up when it was being tested and switch on emission-reducing technology. In normal conditions, the cars roared along with excellent acceleration at hefty speeds, spewing diesel fumes into the atmosphere.

Modern cars containing computer technology often have as much technological sophistication as the average smartphone. They can contain nearly twice the amount of code as you would find in Facebook or the Large Hadron Collider, for example. With such large volumes of code comes a greater risk of vulnerability or error, or potentially deception. They are a good example of how we put vast amounts of computing power into everyday objects and rarely consider the risks associated with allowing the underlying source code to remain secret.

For the last few years, two security researchers, Charlie Miller and Chris Valasek, have taken to hacking various cars, scanning for weaknesses in the code that might allow them to control the vehicle remotely. Miller and Valasek were able to hack into a Ford and a Toyota by "plugging into a diagnostic port that could control the vehicle's steering and speed." As time went on, their hacking gained sophistication, eventually going wireless. Andy Greenberg described driving a Jeep Cherokee as the guinea pig in a troubling experiment:

> I was driving 70 mph on the edge of downtown St. Louis when the exploit began to take hold.
>
> Though I hadn't touched the dashboard, the vents in the Jeep Cherokee

> started blasting cold air at the maximum setting, chilling the sweat on my back through the in-seat climate control system. Next the radio switched to the local hip-hop station and began blaring Skee-lo at full volume. I spun the control knob left and hit the power button, to no avail. Then the windshield wipers turned on, and wiper fluid blurred the glass ...
>
> I mentally congratulated myself on my courage under pressure. That's when they cut the transmission.
>
> Immediately my accelerator stopped working. As I frantically pressed the pedal and watched the RPMs climb, the Jeep lost half its speed, then slowed to a crawl. This occurred just as I reached a long overpass, with no shoulder to offer an escape. The experiment had ceased to be fun.

Hackers in the security field actively look for these weaknesses in software; they are part of the rush to find zero-day vulnerabilities. As discussed earlier in relation to the WannaCry worm, there is a whole market in finding these vulnerabilities and then notifying the company so it can fix them. In many ways, this resembles a much more expensive and inefficient version of a free software project—a roundabout way of getting to the same point. The problem is not just that these vulnerabilities affect software's capacity to do its job properly. Nor is it that the NSA jeopardizes our collective safety (discussed in chapter 3) by accumulating these vulnerabilities as a stockpile of digital weaponry. These vulnerabilities can also be created by companies seeking, for example, to evade legal regulation. Closed source code allows businesses to hack their way around democratically made laws.

Volkswagen's cheating on emissions testing carried on undetected for a considerable period of time, because, just like with the Jeep and most other consumer products that run software, there is no way to read the source code. It is closed source, protected by copyright, and there is no requirement that a company reveal it. It is deeply ironic that the Environmental Protection Agency had actively resisted attempts to exclude this kind of software from copyright protections and make it available for scrutiny. The EPA had argued that making the source code available meant that consumers could potentially tinker with it, to improve the car's performance in violation of environmental regulations. But, unsurprisingly, it turns out that it is not the everyday hacker whom we should fear: businesses are the ones that benefit from this kind of secrecy.

The experts and the media were quick to pillory the company and its conduct as "scandalous," "unusual" with "a degree of complacency and pomposity," an "accident waiting to happen." The company was founded by the Nazis, we were reminded, as though this somehow, however absurdly, explained its bad behavior. But Volkswagen is hardly an outlier. The company's conduct was unlawful, but, on another reading, it was simply exploiting a comparative advantage, externalizing a negative, disrupting a price indicator. Economists are coy about using the proper language for such behavior, which is a form of mundane deception that is the logical incentive of market economics. Why wouldn't companies manipulate software in this way? With their code under the cloak of secrecy, there were minimal prospects of being caught. It is the sensible response to the regulatory intervention overlaid on market forces. "Mundane" now appears to be the most appropriate adjective, given that several other car companies have since come under scrutiny for doing the same thing.

It gives us pause for thought about what might be next on this rickety roller coaster of predatory capitalism. Many consumer products are now reliant on software that is closed source. Expensive medical equipment that falsely reports good results or recommends unnecessary diagnostic testing? Airplanes with bugs in their navigational software that no one noticed, until, say, a plane goes missing? The problem is not merely cheating, which is congruent with the nature of capitalism. The problem is that this kind of reliance on software that is kept secret is, as a matter of design, unsafe. Like a quick paint job before an auction, the cloak of secrecy around software codes conceals the cracks in the foundations. It becomes a dangerous building material.

Free software, with its sprawling and at times chaotic approach to development, has been compared unfavorably to proprietary software, with its sheen of corporate organization and respectability. But one of the key lessons that emerged in the development of Linux—later dubbed "Linus's Law"—is that "given enough eyeballs, all bugs are shallow." In other words, the more people are looking at the code, the quicker its problems are spotted and solved. Open source development—by which I mean projects in which the source is open and available for anyone to see—can be time-consuming and anarchic, but it is also effective. Cheating of the kind done by Volkswagen would be hard to conceal if there were more eyeballs on the code. So too would any design flaws.

A similar approach gave rise to the Agile Manifesto, written by a number of software developers in 2001. Agile management focuses on creating working products quickly and prioritizes cooperation between individuals and teams over formal processes. First regarded as a somewhat hippie approach to organizing production, it has gone on to represent a revolutionary moment in the history of management. Agile is about placing importance on engaging workers (or resisting alienation); it values collaboration (rather than competition); it is about giving people the chance to self-organize (rather than suffer micromanagement). Placing importance on feedback and transparency, often in person, has meant that Agile has become a more accountable form of management that is driven from the bottom up. As a technique, it has transformed software development and is associated with productivity and excellence.

This style of management has its downsides, particularly if implemented thoughtlessly—Facebook's now-discarded mantra of "move fast and break things" comes to mind. It also potentially papers over workplace inequalities with managerial egalitarianism. Similarly, there is a history and tendency among advocates of open source software (as opposed to free-as-in-freedom software) to find ways to make this approach to software production compliant with and complementary to the market. But Agile approaches to both industrial organization and open source design can still produce radical logical outcomes or, as Wendy Liu argues about open source, act as "gateways to a more radical politics" that "pushes for the decommodification of not just information but also the material resources needed to sustain the production of information." These processes and ways of working have the potential to encourage engagement and undermine the exploitative nature of waged work, while building the capacity to generate possible alternatives.

The problem is that computing has become structured in a way that is fundamentally undemocratic. Our whole relationship with personal technology has been organized around the needs of copyright law. That is to say, software has become commodified; the design process serves the interests of the market rather than the interests of the user. This commodification has been the driving force behind the growth of software companies: it locks users into a technological relationship that serves the interests of profit maximization and structures productive capabilities to protect this state of affairs. There is no way for the average consumer to know what their computer, and most of the software on it, is doing. And even if the consumer spots a hazard or a way to do something better, she has no opportunity to communicate this back to a community of users, let alone modify the program. A user can only tell the company and hope they fix it. (Or, less generously, she can profit from companies by selling them zero-day vulnerabilities.) Users are essentially forced onto a one-way superhighway, from producer to consumer, with no opportunity to reverse the direction of engagement. These days, that is not just a matter of getting a program to run on a laptop. As software becomes integrated into more products—as we build the Internet of Things—from cars to fridges to medical equipment and even public infrastructure like mass transit, the potential dangers of keeping this software closed increase.

These are serious and frightening design problems, but they also result in an immense waste of human potential. It is not just that companies do not invite feedback on their software or input from users on their design. Their objective, the purpose of their software, is not to service the user. Their primary goal is to retain control of that software. They want to control who uses it (that is, only paying customers). Proprietary software design makes a fetish of creativity—turning it into something abstract and commodified, geared to the purpose of money-making, rather than a collective or public good.

Proprietary software companies also want to remain the centralized force that dictates how the software develops, which requires the people who use their software to remain ignorant. They want users to feel unable to change their experience of using their product. They want troubleshooting to be something that only a Genius Bar can do. They cultivate a sense of helplessness. This kind of supplicant fatuity ultimately serves their bottom line.

This is not a conspiracy theory. Microsoft developed a corporate strategy that sought to promote this exact psychology in its users. Microsoft generated what it called FUD (fear, uncertainty and doubt) in its customers. It did this in a variety of ways, intended to frighten them into using their software on the assumption that it was more reputable and stable. Such tactics included confected pop-up boxes with red warnings when users sought to download or use competing products. The senior vice president at the time, Brad Silverberg, referred to this strategy in a 1992 email, revealed in a lawsuit against a competitor that sold a product in competition with Microsoft's MS-DOS. "What the [user] is supposed to do is feel uncomfortable," wrote Silverberg, "and when he has bugs, suspect that the problem is [the competing product] and then go out to buy MS-DOS."

Stallman has spoken about how proprietary software seeks to make users "hopelessly dependent" and discourages sharing, with the constant threat of imprisonment. It recalls Marx's description of the experience of labor in the wake of the industrial revolution. The estrangement of labor from production is ultimately one that relies on limiting people's potential.

> Labour produces for the rich wonderful things—but for the worker it produces privation. It produces palaces—but for the worker, hovels. It produces beauty—but for the worker, deformity. It replaces labor

> by machines—but some of the workers it throws back to a barbarous type of labor, and the other workers it turns into machines. It produces intelligence—but for the worker idiocy, cretinism.

Software design in closed environments acts as a brake on human potential in order to sustain the subjugation of technology to commodified form. It keeps users of software benighted, actively denying them the opportunity to self-educate, out of fear as to how this might affect their profitability. It is an enormous squandering of possibility that takes place due to the subordination of software development to shareholder value.

This is not just a shame from a moral perspective. It also represents a power dynamic at play that governs the integrity of our digital systems. Only certain kinds of people think they should get to decide the direction of development of key software programs, including programs that people are dependent on in many aspects of their lives. The public is asked to trust that there is no mistake or deception in their millions of lines of code. The public is also asked to expect that the software they buy will be useful to them, even though it is designed to maximize profit, not utility. Time and again, companies demonstrate that they are not to be trusted and that, without proper regulation, they cheat, undermine their competitors and show disrespect for public resources.

The history of software development provides an insight into some of the biggest barriers to exploring human potential in the future. The proprietary software industry actively sought to resist work being done in an open source format. It blocked paths for collaboration and experimentation; it stood in the way of orienting the collective human brain toward problems people want to solve; instead, it devoted resources to chasing profits. It prefers to risk vulnerabilities, and subsidize the cowboy market in finding and selling them back to companies, rather than open their code to scrutiny. If we apply professor Stafford Beer's maxim that "the purpose of a system is what it does," then what copyright and closed source software do is make money for entrepreneurs using a substandard system of production. The proprietary software industry has produced software that does not serve its users or their collective purposes; rather, it serves the owners of that software.

It is important to imagine a world outside these modes of thinking, because the price we pay for subscribing to the old assumptions is high. With fewer people attempting to solve a problem, seeking creative ways to do something better, or motivated to try something different, the capacity to innovate will be inherently limited. Ada Lovelace was fortunate to have a mother who was determined to teach her math. Without this, she might have been confined to the life of a society woman—and we would have never known as much as we do about the earliest developments in computing. The problem is not that she stepped outside of her socially prescribed role; it is that there were not more people able to do the same.

Limitations in the proprietary software industry may be overcome with an abundance of money and a roomful of highly trained coders. Open source projects are not inherently better; they regularly contain design flaws and often fail to get off the ground as projects. But open source software has better prospects of being effective. Open source software trends toward accountability. Open source software is more likely to service users and come up with new ways to solve problems. Open source software does this because it is serving the people looking at and using the code, not the owners of a corporation, in a manner that is transparent. The best kind of regulation comes through transparency, through which all bugs become shallow. Proprietary software puts profit before safety, efficiency and the public welfare. Open source works because it allows the sunshine to kill all the bugs.

Collaborative innovation means more people working together to find and solve problems and more people benefitting from a better program. It prioritizes the integrity of the product over the need to make money and resists the shoddy workmanship that comes with market incentives. It involves a diversity of viewpoints and a variety of skills applied to the necessary tasks, encouraging us to rethink traditional understandings of how innovation happens, which have tended to lionize individual genius. It encourages us see how a motley range of people contributing to poetical science and poetical philosophy can generate technological progress. Human beings often do their best work when they are able to produce themselves intellectually but also actively, and, in doing so, contemplate themselves in a world they have created.

We need to make the open source software movement dangerous again. The free software movement attracted the wrath of all the right people, it created the possibility of building technology differently, and we need to reproduce the radical ways in which that movement approached the idea of work and objectives of production. This will require us to cultivate a diverse and inclusive workforce making open source software whose contributions will not be exploited but properly valued and supported as a contribution to the common good. This is especially important as many people who make software are not doing so in conditions of freedom or even in gentrified settings like Silicon Valley. Producers of software are working for pay in all sorts of places, from Kerala to Shenzhan, as well as often contributing to open source projects in their spare time. If we can find common ground in an understanding of the limits of proprietary software and the potential of free software, we have the potential to organize a whole new cohort of workers around a goal that challenges some of the key tenets of capitalism.

The gods treated Babbage cruelly by, as one historian put it, "having bestowed upon him a vision of a computer, without granting him the tools—technological, financial and diplomatic—to make his dreams come true." We now inhabit a world where Babbage's dreams are being realized, and Lovelace's logic has taken root in the minds of many. Their experiments combining poetry and mathematics, imagination and practicality, were the beginning of a field that is changing our society in all sorts of radical ways. But there are limits on this kind of collaborative human potential, and they are ones that are imposed by capitalism. We need to work toward creating conditions in which those limits no longer apply. In imagining what this might look like, we need not begin from a blank slate or concocted utopias, severed from the past. Here, in this specific history of the industry, we see the seeds of alternative ways of working that are fulfiling and liberating; we glimpse the possibilities of production outside the marketplace.

7

Digital Citizenship Is a Collective Endeavor

Tom Paine's Revolutionary Idea of Public Participation

The first half of Tom Paine's life was characterized by "unrelenting failure," according to the historian Eric Foner. Born in 1737, Paine grew up in an England defined by mass poverty. He managed to lose jobs, his livelihood and even two wives over the first decade and a half of his adulthood. What he did gain, however, was some experience in struggle. While working as a tax collector, he led a campaign for better wages. Naturally, it too ended in defeat.

This episode left him with a strong empathy for and identification with poor and working people. It was a political orientation that he carried with him to America. He traveled across the Atlantic in 1774, with a desire to start afresh and a vision of a secular and free society. The America he found, however, was stifled by colonial rule. Frustrated but determined, Paine turned to writing to formulate and spread his ideas. "When I reflect on the horrid cruelties exercised by Britain in the East Indies ... When I read of the wretched natives being blown away, for no other crime than because, sickened with the miserable scene, they refused to fight," wrote Paine in 1775, in anticipation of events that would later come to pass, "I firmly believe that the Almighty, in compassion to mankind, will curtail the power of Britain." Shortly thereafter, Paine was involved in the

American Revolution, a grand and dramatic curtailment of British power meted out not by God but by the people.

Paine possessed a talent for incendiary and inspiring rhetoric, and he was the first writer to frame the possibility of a republic in virtuous rather than derogatory terms. He showed radical indifference to the supposed superiority of hereditary nobility and lacked the distaste for the poor and disenfranchised that characterized so much political thinking at the time. His empathy with working people ran deep: he understood that a system of government run by common people, for common people, was a worthwhile pursuit. Written just a year after he arrived in America, Paine's *Common Sense* became one of the most popular and influential political pamphlets of all time. It is commonly credited as a first draft of the Declaration of Independence. The practical experience of arguing against privilege and in favor of popular government was part of Paine's lifelong commitment to these ideas. This led to his involvement in the French Revolution, an event that took many of these ideas even further.

As a polemicist, he was well-suited to his time, but Paine the philosopher was far ahead of his contemporaries. This was particularly true in one key respect: he was one of the earliest advocates for a comprehensive system of social welfare. As early as 1775, he talked about his "plan for raising a fund for the purpose of portioning off young married people, with a reasonable sufficiency to begin the world … and likewise for raising another fund for the purpose of supporting us in our old age." This was not about pity, or charity. For Paine, provision for people at both the beginning and end of their productive lives was "a necessary and valuable appendage to our present circumstances."

Toward the end of his life, Paine returned to these ideas more comprehensively in his pamphlet *Agrarian Justice* (1796). The central tenet of his argument was that every person was entitled to some share of commonly held resources:

> The earth, in its natural uncultivated state was, and ever would have continued to be, *the common property of the human race* … Every proprietor,

> therefore, of cultivated land, owes to the community a *groundrent* (for I know of no better term to express the idea) for the land which he holds; and it is from this groundrent that the fund proposed in this plan is to issue.

Paine acknowledged that cultivation of the land increased its productive capacity, creating some kind of entitlement to the proceeds on the part of those who worked to achieve it. But he argued that this does not extinguish the common claim that humans collectively hold to share in the natural wealth of the world. In other words, people have an entitlement over what society produces not because of the virtuousness of their actions but as a function of their status as human beings.

While this is often referred to as the earliest conception of a universal basic income, Paine's proposal is better understood as a universal capital grant. A universal basic income is a regular payment made to all citizens without qualification. It is a redistributive measure that aims to give everyone a substantively equal share in the productive output of society. A universal capital grant, on the other hand, aspires to create equality of opportunity, through payment of a lump sum at a particular age (say, eighteen years). The idea held special relevance in a society like Paine's, where land was the source of significant wealth and seen as a ticket to social mobility. A universal capital grant could allow young adults to purchase land, as part of an active effort to level the playing field, to allow talent the opportunity to flourish and afford hard work its due recognition. It was an attempt to redress the inequality that came about through land ownership structures developed to preserve the privileges of the aristocracy.

But it was also an acknowledgement that prosperity from land cultivation was not an individual pursuit but rather occurred in a social context. To cultivate land, you needed a marketplace for exchanging goods, as well as physical infrastructure such as roads and civil systems such as law, which were all organized collectively. The payment of a ground rent was the price individuals paid to the community to exploit land for their personal gain and make use of the common resources that allowed them to do so. It was also a radical political intervention into a society that sought to abolish notions of aristocratic privilege, with an argument for a just redistribution of resources.

Paine's writings mark the beginning in Western political thought of a very long tradition, namely that people are worth more than their productive capacity. People are not simply economic units whose moral worth is measured by what they produce for the market. Every person is entitled to a share in the wealth that is created collectively in the world, as part of recognizing his or her existence. The capacity to feed, clothe and shelter oneself ought not be dependent on the capacity to do some kind of productive work, especially when being productive is defined within the narrow confines of material value determined by market capitalism. The moral hierarchy created around people's economic success as a productive unit is false and dehumanizing. People have a claim upon society by virtue of their humanity.

The other side of this philosophical coin is marked by the recognition that inequality undermines democratic structures. This is easy to miss if we understand democracy in purely formal, representative terms. The fledgling democracies of America and France were exciting because they threw off the stuffy notions of aristocratic privilege. But inequality risked undermining the freedom that was being explored in these new societies. "The peer and the beggar are often of the same family," observed Paine. "One extreme produces the other: to make one rich many must be made poor; neither can the system be supported by other means." Poverty and discontent and the social outrage generated by this combination shows that "something is wrong in the system of government that injures the felicity by which society is to be preserved." Foner argues that, for Paine, "poverty was a product of civilization, not nature"—a radical position in an age that thought of itself as meritocratic and rewarding of hard work. Paine never took this argument further, as others did, to extend his critique to property in general as the source of inequality.

But no one could be in any doubt of his feelings on the subject of poverty. For Paine, "the present state of civilization is as odious as it is unjust."

The American and French Revolutions introduced a new and radical way of ordering society. Those moments had a global impact, including in countries like Haiti, where the ideals of liberty and equality were extended by revolutionary leaders to attack the institution of slavery. But this new order contained all sorts of problems and contradictions. Criticisms of the nature and limits of representative democracy remain relevant centuries later and have spurred radical imaginings ever since. It is easy, as many modern reactionaries and conservatives have done, to cast Paine as a libertarian, concerned above all with the excesses and injustices of government. But this would be like interpreting the American Revolution as nothing more than a tax revolt: a narrow and reactionary reading of history by those determined to ignore the alternative possibilities germinating among the grassroots. One of the central ideas in the tradition that Paine contributed to affirms quite the opposite: we need more democracy, not less. In a truly democratic society, people should also have the opportunity to deliberate on economic questions, to discuss how value is generated and to whom it should be distributed. These are not simply private matters to be left to the market.

What does this political tradition mean in positive, practical terms in the wake of the digital revolution? This is the subject of this chapter and the next.

In this chapter, I want to discuss the importance of access to the network for the purposes of participating in society, to engage equally in the public spaces and platforms that facilitate democratic decision-making in the modern age. Formal equality in the digital age—the right to access rights via the network—is an important aim. I then talk about how digital technology can facilitate the redistribution of wealth and democratize production, and labor is central to these pursuits. Because formal equality, or networked enfranchisement, is nothing without substantive equality too.

In the first half of this book, we have looked at some of the oppressive qualities of the digital revolution—corporate and state surveillance, algorithmic bias, technological utopianism and the commodification of innovation and software production—and how we might be able to resist these. But networked technology is also full of potential to experiment with improved forms of democracy and social organization, including how we understand citizenship, our relationship to work and one another, and the stewardship of collective resources. In the second half of this book, I want to consider the possibilities for greater democracy and dignity and the potential of networked technology to create the conditions necessary for us to win this world.

When Facebook created its platform Free Basics, it epitomized a kind of mafia capitalism: an offer too good for poor people to refuse. Free Basics is what is known as a zero-rated service. It involves an agreement by Internet service providers and cell phone networks to not charge customers for data when using certain services or websites, allowing people to get onto certain parts of the web at very little or no cost. The idea behind such services is that the Internet is good for the world and even better when more people are on it. The ostensible aim of Free Basics is to overcome the digital divide between those who are online and those who are not. Free Basics includes a number of websites, among them Facebook, of course, and is currently offered in forty-seven countries. According to the company, it is responsible for bringing 25 million people online.

One country that faces a significant digital divide—and the associated social and political problems that underly it—is the world's second most populous nation, India. Around 67 percent of India's population live in rural areas (compared to 18 percent of Americans, for example). These people, in particular, encounter barriers to getting online, from the cost of data to the quality of infrastructure.

Mark Zuckerberg, Facebook's (in)famous CEO, spent some time in India in 2015, eager to win people to the idea of Free Basics. In a post for the *Times of India*, Zuckerberg listed all the amazing things that can be achieved by connecting people to the Internet, from ending poverty to improving education and job opportunities. Zuckerberg claimed that "the biggest barriers to connecting people are affordability and awareness of the Internet." In later speeches, he put this in even more specific terms: "The biggest reason why four billion people don't access the Internet—bigger than issues of access or cost—is that they don't know why it might be valuable."

Rest assured, people of the world, Free Basics was there to fix the biggest reason for the digital divide! "Who could possibly be against this?" asked Zuckerberg, presumably rhetorically, as the actual answer was not one that he wanted to hear. In early 2016, the relevant regulatory authority banned Free Basics and other zero-rated services in India.

Zuckerberg's pitch was based on the idea that the digital divide arises from ignorance and fear of the unknown among the great unwashed. Skepticism toward his project was, by association, perceived as backward. By Zuckerberg's reasoning, any forward thinker would embrace the expansion of capitalism and the economic development that would flow in its wake. It all had a bit of a Rudyard Kipling feel to it, as though the enlightened Zuckerberg was taking up his burden to bring the future to a nation that was captive to the past.

Unfortunately, the truth is a little less grandiose. It turns out that most people using Free Basics are probably *not* first-time Internet users, or, as Zuckerberg might put it, people who have finally learned that access to the Internet might be valuable. On the contrary, according to one journalist, Free Basics is basically "a winning customer acquisition strategy" for networks. Many carrier services are happy to waive the cost of data on Free Basics because it is in their interests to do so (note that it is these companies, not Facebook, that actually foot the bill for zero-rated services). This is because many customers use Free Basics when they run out of credit. Rather than being a way to get first-time users onto the Internet, Free Basics is a way for telecommunication companies to retain customers, even giving them a competitive edge over their rivals.

For Facebook, it is a vital way to continue growing user numbers, which is a central plank of their growth strategy. Many Indians rapidly saw through the scheme: "Their pitch about access turned into mobilization for their own product," said Osama Manzar of the Digital Empowerment Foundation. Indians were unimpressed at being patronized, and an enormous campaign against Free Basics within India emerged as a result.

The more humdrum problems of poverty and governance are arguably greater contributors to the digital divide than the benighted views of poor people. Access and cost of data in India are major problems. Only about a third of Indians are actually connected to the Internet. Rural Indians are underrepresented in connectivity terms: despite making up over two-thirds of the general population, only 15 percent of rural Indians are Internet subscribers. Infrastructure, more than individual or cultural attitudes, is the key to improving these numbers.

It was not long before the disappointment at the Indian regulator's decision revealed the rather crude politics behind Facebook's lofty rhetoric around Free Basics. India was making its own decisions, as a sovereign nation, without proper deference to the interests of Western business. In response, the Facebook board member and famed venture capitalist Marc Andreessen tweeted: "Anti-colonialism has been economically catastrophic for the Indian people for decades. Why stop now?" Andreessen quickly deleted the offending tweet, and Zuckerberg distanced himself from Andreessen's seeming soft spot for the days of the British Raj. Andreessen apologized.

But this was not simply a problem of optics, ignorance, or careless social media use. In an analysis of the saga, Rajat Agrawal pointed out that Facebook might have been wiser to focus on giving not "unlimited access to a restricted Internet" but "limited but free access to the entire Internet." Indeed, this was the exact strategy of one of Facebook's key rivals, Google.

In 2015, as Zuckerberg was making his ill-fated attempts to woo Indian regulators, Google announced a new project to provide high-speed public Wi-Fi in 400 train stations across India. "Even with just the first 100 stations online, this project will make Wi-Fi available for the more than 10 million people who pass through every day," wrote Google's CEO, Sundar Pichai. "This will rank it as the largest public Wi-Fi project in India, and among the largest in the world, by number of potential users." Google was focusing not on giving lots of poor people a slice of the Internet for free, but building infrastructure that could give anyone nearby, be they rich or poor, the entire Internet for free.

Left out of the glossy blog post and gushing media coverage were some other realities of these proposals. This was a large-scale public infrastructure project that was handed over almost wholesale to a private company, without a transparent tender process. And just because the service is free does not mean that no one is paying a price. Though this service, Google will be able to hoover up vast amounts of data from people joining its network. As Sreejith Panickar put it: "Google got the monopoly of the service without having to bid for it. In effect, the free broadband service was the green flag Google waved to make its entry to our railway stations." Panickar predicted that this free access could well end up as a paying service at some point in the future. The company will find a way to make a return on their investment, he argued, through monetizing data or otherwise.

A very similar program to Google's Indian Wi-Fi project has also been rolled out in New York City. This project ultimately aims to replace the city's pay phones with 7,500 street kiosks, offering super-fast Internet connections for free. As in India, these kiosks allow Google to collect hefty swathes of users' personal data, often without them realizing it. What ought to have been a public infrastructure program was handed over to a private company with little oversight on behalf of consumers, let alone a formal tender or bidding process. The outcome is "a real-time, personalized propaganda engine," says author Douglas Rushkoff. These are public infrastructure projects built by private enterprise in exchange for a monopoly over data collection. Residents do not have to pay for them, but there is a price. They will be funded almost entirely through advertising.

The attempts by tech giants to tackle the digital divide tell us several things. Digital platforms are racing to claim market share, and to that end they are prepared to spend money on projects that might otherwise have been carried out at public expense. We should be wary of this kind of private incursion into the public domain and the antidemocratic power dynamic that it creates. It is a gift horse that needs a thorough dental examination. Free Basics in India was not an example of Facebook having good, albeit poorly thought through, intentions. The alternative put forward by Google was not an example of that company devoting its skills more effectively to a philanthropic project in the developing world. The whole affair was a face-off between rivals in the race to digitally colonize the world. These platforms aim to dominate the public space, to be the controlling platform for the marketplace for advertising traditional goods and services. They seek to render the state, and public spending projects, unnecessary. This has very serious consequences, not least that these companies also become the controlling platform for the distribution of ideas.

Bridging the digital divide is an important project for the billions of people who want to get online and participate in society using digital technology. Just like many people struggle to access clean drinking water, electricity or basic plumbing, access to the Internet and computing is best thought of as a kind of utility. Ensuring public participation in the digital age will mean spending public funds on infrastructure projects, transparently designed and implemented, with the common good as the central objective. Infrastructure that provides access to the Internet is an essential collective resource and must be funded publicly. It should not be something that you are made to pay for individually, either in dollars or in data.

This idea could be called the "right to the Internet," but that is too discrete and narrow in its expression to give full weight to the idea—the ability to read, to write, to speak and learn are bound up with this right. The phrase also implies that the Internet is a single entity, an immutable and neutral network, when the reality is less straightforward. The material access to these rights—to the capacity to exercise them—is more complex than the rights themselves.

Networks vary in quality. By now, the words "network neutrality" have made their way out of the lexicon of nerds and policy wonks and into mainstream parlance. But allow me to explain: the term is about applying "common carrier" rules to the Internet. That is, a system whereby carriers of traffic on the network do not discriminate by content. All telegrams were historically delivered at the same pace, regardless of their destination or what they contained; power comes out of the outlets, no matter which appliance is plugged in. The principle of net neutrality requires that information traveling through the network not be tampered with by, say, slowing it down according to the preferences or interests of the provider. Network neutrality is the principle that all Internet service providers should be neutral in the service they provide: they should favor no traffic or application on grounds of source or content.

When professor Tim Wu first coined the term "net(work) neutrality" in 2003, the central justification for his argument in its favor was the principle of innovation. "A communications network like the Internet can be seen as a platform for a competition among application developers," he wrote. "It is therefore important that the platform be neutral to ensure the competition remains meritocratic." In Wu's vision, market capitalism would lead to innovation because a neutral network would allow good products to take the place of bad ones. It would mean that companies cannot limit access to other, competing content by flexing their business muscle. It was a way of avoiding monopolistic tendencies when it came to content creation.

Net neutrality has been dealt a serious blow in the United States under President Trump's administration, though the fight is not over. Protecting it remains a priority for activists all over the world. But a neutral network that respects common carrier rules, while essential, is not the same as a democratic network. Net neutrality is not the only priority for activists, because Internet service providers are not the only entities capable of shaping the network in their own interests. To put it in terms that Paine might understand:

digital technology creates a form of enfranchisement of everyday people; it provides the possibility that they can have a greater say in matters of public policy and enhanced accountability over those who govern. But it requires us to also build the structures that actively protect the democratic nature of this space. Access to the public utility we know of as the Internet is the method by which we access many other rights we have in modern society, but that access is subject to all sorts of organizational discipline and gatekeeping, which undermines its capacity to empower the citizenry. Networked enfranchisement will require that we maintain the infrastructure of this right—accountability over the pipes and switches that facilitate collective decision-making.

The right to vote in revolutionary America was a central demand in the fight to break the chains of colonial power. The right to elect a representative was "the primary right by which other rights are protected," Paine wrote. "To take away this right is to reduce a man to slavery." But it is important to understand this right in context, not merely as a formal right. It is a "right to have rights," as Hannah Arendt might have put it, or a right we "cannot not want," as Gayatri Spivak might say. The right to vote is essential to any form of public participation, but it is not enough. While it is something that we must defend, this is a two-faced task, because we also must maintain an understanding of its limitations. The right to vote loses all meaning if it is defined narrowly; we need to defend it as part of a project to reclaim the expansive possibilities of public participation.

Over ten years after he coined the term "net neutrality," Wu was openly critical about the effects of the commodification of the Internet. "The business model demanded its payout ... The web has gotten worse over the last five years, as opposed to better." Wu pointed out that Google devotes an increasing amount of space within its search results to advertising, with design attributes that make them more and more difficult to distinguish from organic results. We have predictability instead of serendipity; conformity replaces individuality; the interests of a few are prioritized over empathy with the many.

Just like the wealthy aristocrats and benighted clergy who presided over the society Paine was part of overthrowing, technology capitalists have outlived their usefulness and are undermining our advances to protect their outdated sense of privilege. Common carrier rules are important, but we must also turn a critical eye to other layers of the network.

In the digital age, the Internet is increasingly the place in which rights gain meaning. We still vote at a physical ballot box, but the public spaces where we learn about how we are governed and engage in public discussion are online. To fail to make these digital spaces available to all citizens is akin to a failure to provide them with basic literacy skills. Access to the Internet is a basic requirement for public participation. But equally, to allow gatekeepers like Facebook or Google to place themselves between individuals and the public space creates a power dynamic that is profoundly damaging.

When power is concentrated in large private organizations, particularly digital platforms, political censorship can be effected with ease. Facebook has admitted that it deletes accounts in response to requests from the Israeli and US governments. Google provides significant funding to the New America Foundation, which infamously fired an employee that was critical of the company's conduct. LinkedIn was the most explicit, when discussing its entry into China. After a backlash in relation to censorship of content relating to the anniversary of the Tiananmen Square massacre, a spokesperson was blunt: "It's clear to us that in order to create value for our members in China and around the world, we will need to implement the Chinese government's restrictions on content, when and to the extent required." As Glenn Greenwald has observed, these companies, in the same way that governments do, "will use their censorship power to serve, not to undermine, the world's most powerful factions."

Platform monopolies diminish the experience of being in the public space in more subtle ways too. Particularly in the wake of the 2016 US presidential election, it has been common to blame the public for choosing to live out much of their lives in the fragmented communities of social media. We are told we only listen to people we agree with, that we live in filter bubbles and echo chambers, or that we are hopelessly naive and influenced by fake news. But this ignores many contextual problems, including how the dominant platforms have a structured predilection for polarization. Facebook's objective is to make people stay on the site, and its algorithms are designed to use our emotions to do this. It curates the news feed to display stories that confirm the political views of users, thereby keeping them engaged. The data that these companies collect become a lens that can perform a subtle kind of organizing without our knowledge or consent.

Facebook does more than just tinker around with content, it actively manipulates. Most infamously, Facebook ran experiments on its user base in 2014. It manipulated the news feed to show more or less positive news stories, in two parallel experiments, without the knowledge of users. The results were published by the Facebook Core Data Science Team:

> Emotional states can be transferred to others via emotional contagion, leading people to experience the same emotions without their awareness ... When positive expressions were reduced, people produced fewer positive posts and more negative posts; when negative expressions were reduced, the opposite pattern occurred. These results indicate that emotions expressed by others on Facebook influence our own emotions, constituting experimental evidence for massive-scale contagion via social networks.

There was immediate uproar about the ethics of this experiment, which involved 689,000 users without their knowledge or consent. Commentators raised concerns about the possible manipulation by Facebook for political gain—concerns which now appear prophetic. Mike Schroepfer, Facebook's chief technology officer, later admitted the company was "unprepared for the reaction the paper received." He pledged to take steps to avoid repeating the error, including establishing guidelines for such research going forward.

After the election of President Trump, this debate gained a whole new momentum. Concerns about the polarizing nature of social media were rife, and there was a growing sense that a platform like Facebook needed to take responsibility for how this phenomenon was playing out, even if what that might look like remained unclear. When it emerged in 2018 that Facebook had allowed Cambridge Analytica to harvest huge amounts of user data to inform its strategizing on behalf of political groups, for the first time the company faced an unprecedented public and political backlash. Facebook itself may not be directly attempting to manipulate users, but the company created the infrastructure for others to make their best efforts to do so. It created a profitable platform that facilitated the "democratization of propaganda" and the amplification of division. It is not yet clear the extent to which these new forms of information dissemination impact voting patterns, but we can confidently say that the dysfunctional nature of social democracies is not simply attributable to ignorance or bigotry on the part of voters. There are larger forces at work.

Technology capitalism, for the most part, treats this polarization in contradictory ways. Companies profit off the back of claims about their capacity to target specific audiences, but they also appear baffled by the extremism or divisiveness driven by the products they create. Social media platforms benefit financially from creating "a dynamic where people are pulled to the extremes," as Maciej Cegłowski observed. As a *BuzzFeed* report on this issue concluded, "the best way to attract and grow an audience for political content on the world's biggest social network is to eschew factual reporting and instead play to partisan biases using false or misleading information that simply tells people what they want to hear." The fragmentation of online communities and polarization of public discourse is an outcome of Facebook's desire to monetize the web through optimally crafted audiences for advertising. Now that the public are increasingly aware of this business model and object to it, however, its sustainability would appear to be compromised.

It is important to remember that digital behemoths like Facebook and Google are not public service providers. Google is not a search engine company, or an email company, or a mapping service. It is an advertising company. Alphabet, Google's parent company, made 84 percent of its revenue from advertising in 2018. Facebook is even more dependent on advertising: it generated around 98 percent of takings from ad revenue during this time. Outside of China, Google and Facebook account for 84 percent of the global digital ad market; together they account for around 58 percent of online ad spending in America. Of course, they also provide other services and have a motivation to make products that serve the interests of users, but their customer base is quite different to their user base. These two companies are creating an oligarchy in the market for advertising, and it shapes everything that they do.

Technology capitalists' pose of innocent surprise before the degenerative, antidemocratic nature of online spaces is wearing decidedly thin. There is mounting pressure for companies like Facebook to take greater responsibility for the wider impact of their business decisions, both socially and electorally. An early investor in Facebook describes the history of the company as resembling "the plot of a sci-fi novel: a technology celebrated for bringing people together is exploited by a hostile power to drive people apart, undermine democracy, and create misery." He is just one of many former senior staff to criticize the company. It is reasonable to treat these regretful executives with skepticism—they are, after all, the same people who profited nicely from the business model when it was getting started. But their growing number symbolizes something important. It is an opportunity to explore how digital technology might better serve the democracy we live under (however flawed and limited in its current form) and think more broadly about how networked technology could facilitate more meaningful forms of public participation.

A curated, polarized news feed undermines our sense of what it is to be a member of the public. It creates a mediated experience for users; it makes us members of a rigged network. The profit motive of Facebook and other companies degrades the openness of our experience on the web and undermines equality. Digital technology makes it possible to connect across space, class and culture and build resilient and sophisticated communities, without the usual institutional gatekeepers standing in the way. It has the capacity to create space for a universal, international republic, which transcends national borders and social constructs like race and gender. But rather than humanizing people who are different from us, one of the main platforms through which people exchange ideas is doing the exact opposite. It is collecting data about us to organize us into categories; it is designing platforms that reinforce those categories and encourage polarization and hyper-partisanship.

These kinds of gatekeepers are not new, and we should be cautious about overestimating their capacity to exercise power over us. Traditional newspapers created a mediated experience too, in the sense that editors chose what stories to run and where. But the scale is new. Facebook, Google and Amazon wield significant amounts of power over our experience as consumers but also over what we read and discuss as public citizens. Around 60 percent of Americans get their news from social media, and 66 percent of Facebook users get their news on the site. Google/Alphabet has a fully integrated vertical monopoly on advertising, from placement to data analytics. Amazon dominates the market for online book sales, controlling two-thirds of online print and e-book sales, has a growing share of the online ad spending, and has also insinuated itself into the publishing market, inching toward a vertical monopoly. These platforms curate content for readerships that are larger than any audience enjoyed by any other publishers in history. They are creating their own idea of the public and the delivery of public services, while eroding the sense of a universal experience, undermining the sense that such a thing is possible. Formerly, the idea of a public audience was equally curated by traditional media like television and newspapers. This had its problems. But there was an element of transparency: readers or viewers all saw the same material and the same adverts. Those days are over, but we are yet to develop the tools and language to resist the tendency of technology capitalism to undermine democracy.

In the past this kind of dominance was often addressed by competition law, which was alert to the political implications of monopolization. Legal academic Lina Ahmed argues that the importance of antitrust law has traditionally not just been economic, it was also understood in political terms. The vision of legislators was animated by an understanding that the "concentration of economic power also consolidates political power." Monopolies that vest control of markets in a single person create "a kingly prerogative, inconsistent with our form of government," declared Senator John Sherman in 1890 when introducing the Sherman Act, the key statute in American antitrust law. "If anything is wrong this is wrong. If we will not endure a king as a political power we should not endure a king over the production, transportation, and sale of any of the necessaries of life." But since the 1970s, antitrust law has prioritized consumer welfare—often understood in the form of lower prices—above all other concerns. This means that platform monopolies defy convention when examined through the lens of antitrust objections to monopolies.

The common warning about communism was that we would all have to buy goods from a state supplier—yet modern platform capitalism now means that we have to buy everything from Amazon. These companies end up controlling large sections of infrastructure that ought to be under public control. These monopolies—which now cover Internet connectivity, logistics, book publishing and news dissemination via a monopoly in advertising—limit the goods available in the market, but they also curtail our capacity to access public spaces and communities.

It is time these enormous companies were broken up and the vertical monopolies they have created were prohibited. To the extent that technology platforms provide services for public benefit—such as communication, logistics or server hosting, to name a few—they should be held to a standard of nondiscrimination in terms of service delivery and prevented from using this position to build and consolidate other aspects of their business. These companies are increasingly providing the utilities of the digital age—the infrastructure through which we exercise our rights—and it is too dangerous to leave them in the control of private interests.

A democratic society requires not only individual rights but also a shared sense of what it is to be a member of the public. This creates a forum for collectively exchanging ideas and experiences. Participatory democracy is an essential part of justice. The political theorist Nancy Fraser defines justice to mean "social arrangements that permit all to participate as peers in social life." This sense of the universal—of a public set of common values and aspirations—also gives meaning to individual rights. It translates the experience of the personal into a common and public one, and how we frame what is important to us collectively will in turn be refracted back into our everyday experiences. Digital technology, with its capacity to instantly connect across space, create communities and share knowledge from the most disparate sources, presents one of the greatest opportunities to achieve this aim. But technology capitalism is busy fragmenting this, modeling the public around private interests. "When the rich plunder the poor of his rights," warned Paine, "it becomes an example to the poor to plunder the rich of his property."

Tom Paine was a failed organizer himself, but he understood the importance of organizing working people as part of the struggle for change. In some ways, we are starting to see similar activities within the technology industry today. Tech workers all over the United States, as well as in countries like Brazil or India, are organizing to demand better workplace rights. Workers within this industry are some of the best-placed people to offer a critical account of how their companies work. They have much to teach us, not just about the limits and problems internal to the industry but also about the impacts of technology on democracy. "Humanity faces actual existential threats right now, like massive inequality or global warming, and these guys aren't going to innovate us out of them," argues Ares

Geovanos of the Tech Workers Coalition, an organization of tech workers and activists based in the United States.

The idea of democracy requires us to think beyond the ballot box, to see voting as just one point in a constellation of engagements that make up the public space. Public discussions and public life increasingly take place online, and, as such, both the physical infrastructure of the network and the coding of the web must renounce discriminatory practices. We need public spaces that are transparent, accountable to the public interest and resistant to the commodification of users, so that the debates that are vital and necessary can happen. Our public spaces, as they exist online, are not necessarily neutral or equal, but they can be, and we must demand that they be made so.

Despite his success as a pamphleteer, Paine's relentless criticism of organized religion eventually alienated him from the Christian fervor gripping the United States in his final years. Thomas Jefferson, an erstwhile political ally in the American Revolution, distanced himself from Paine when his religion became an electoral issue. Even thirteen years after Paine's death, Jefferson rejected a request for a letter of his addressed to Paine to appear in print. "No, my dear sir, not for this world," he replied. "Into what a hornet's nest would it thrust my head!"

Paine died in 1809. Only a handful of people attended his funeral—two of whom were African Americans, their presence attesting to Paine's lifelong commitment to the abolition of slavery. For the most part, his death went unnoticed in the American press. It was not until years later that some began to hold commemorations of this man's remarkable life. Who were the first to undertake this task? People who were associated with the emergence of the class-conscious labor movement in America. Paine's philosophical legacy belongs where his political life began, among the poor and working classes who struggle to build a truly democratic society in which wealth is shared among the many, not the few.

8

Automation Can Mean Less Work and More Living

Downing Tools so We Can Build Robots to Eat the Rich

In the foundations of the old law quadrangle at the University of Melbourne there is a plaque. The quad was built in 1854, and the plaque was placed two years later. It commemorates the historic strike by stonemasons on-site. They had downed tools and marched to Parliament, protesting their recalcitrant contractor who was not respecting the length of the workday that had been negotiated. After an inquiry, the government officially sanctioned their demand for an eight-hour day, paving the way for the system to spread to other industries.

The plaque is inconspicuous, and the demand hardly seems radical by today's standards, but it was the first institutional recognition of a movement that challenged the power of capital all over the world—the movement that brought us the weekend and the eight-hour work day. "[This event] was one of the earliest to establish an officially sanctioned standard at leading sites for a whole industry across a specific region," writes historian Peter Love. "All the main participants fully understood the significance of what had happened and marked the occasion with a celebratory dinner." That dinner marked the beginning of the tradition of working-class celebrations to commemorate the eight-hour day—which would eventually come to be known as May Day.

The downing of tools in the Antipodes was vitally connected to a large transverse movement around the globe for shorter working hours. It started as a campaign to reduce the working hours of children in England. The Cotton Mills Act passed in 1819 prohibited children under the age of nine from working in cotton mills and limited the work of people under sixteen to twelve hours a day. From here, the campaign for shorter hours for all workers gained traction globally: in 1835, there was a general strike in Philadelphia that won the ten-hour day. A little over a decade later in England, the Factories Act of 1847 mandated a ten-hour day for women and children. Following in the footsteps of the workers of Melbourne in 1854, the campaign for an eight-hour day increased in popularity, with May Day marking an annual international call for the legal protection of the eight-hour day from as early as 1871.

The eight-hour-day movement became popular with workers all around the globe. It was one of the most successful progressive social movements in history. It was internationalist in spirit, attracting support from millions across several continents, and the demand was ultimately enshrined in law in multiple countries. This represents a monumental achievement by almost any standards.

In many ways, the workers who fought for limits to the length of the working day were grappling with problems similar to those that confront society today. Nineteenth-century industrialization transformed the workday in a rapid and unprecedented manner. From a routine of labor that was built around natural limitations, such as seasons, weather and daylight, the industrial revolution brought forth the hellish possibility of ceaseless production. Industrial machinery could run continuously, meaning that the limits of the workday were defined by the physical capacity of human bodies. The time that work finished was determined by their exhaustion, as though people were nothing more than an economic unit. "Human endurance was for many years the sole check upon a day's labor," wrote the economic historian Frank McVey in 1902, reflecting on this situation in his polemic in support of the eight-hour day. "In these days of industrial concentration and wealth-getting," he wrote, "the whole industrial machinery is organized for the mere sake of production and the profit incidental thereto. Workers, under such a conception, are regarded as the parts of a machine system instead of members of a society."

Perhaps most famously, the horror show of the industrial revolution for working people was forcefully put on display by Friedrich Engels in *The Conditions of the Working Class in England* (1845). In this careful and detailed work, Engels argued that the economic drivers and the health effects of industrialization were such that working people were worse off, even as society was more productive and technologically advanced than it had been a generation or two earlier. "The manner in which the great multitude of the poor is treated by society today is revolting," wrote Engels. "How is it possible, under such conditions, for the lower class to be healthy and long lived? What else can be expected than an excessive mortality, an unbroken series of epidemics, a progressive deterioration in the physique of the working population?"

In something of a future echo of Engels's observations, the development of digital technology has correlated with a range of poor outcomes for working people. The rapid development of digital technology in the twenty-first century has meant that work has broken out of its "temporal and spatial confinements." People in many parts of the world are almost continuously on the job, in the office, on the road or at home, constantly connected to devices. For other workers, the piecework economy has created a precarious existence of short-term gigs and long-term worries.

These trends have coincided with increasingly economic inequality. Wealth inequality is the highest it has been in most countries in thirty years. Inequality of income (annual wages) in a place like America is obscene: the top 10 percent of households receive 28 percent of income. But this at least remains relatively similar to other countries. On the other hand, inequality of wealth (total assets minus liabilities) is more insidious and troubling: the top 10 percent of American households own 76 percent of wealth. Interestingly, health indicators are also declining: life expectancy has stalled or even receded for certain sectors of the American and British populations. Death rates among white middle-aged people are on the rise, as a result of things like suicide and increased substance abuse. This is significant because it contrasts with every other age group, every other racial and ethnic group, and equivalent groups in other wealthy countries. Though there is no clear reason for this trend, it seems to be a function of physical pain, financial distress and mental illness.

Meanwhile, millions of people are laboring in conditions that perhaps best reflect the modern version of Engels's tract in places like China that host the lion's share of manufacturing of digital hardware. These people work in factories that are often brutal and at times deadly. Production takes place around the clock, at inhuman speeds, for low pay. Fatal explosions are common. Jack Qiu makes a compelling case that factories like those managed by Foxconn practice a form of slavery. Each factory is run like an "independent kingdom," violence toward workers by private guards is common, and workers have no recourse to any public authority. To leave the job involves clearing immense bureaucratic hurdles, at the risk of forgoing owed wages. In 2010, this situation led to what has been called the "Foxconn Suicide Express": fifteen workers tried to kill themselves over a period of five months by jumping from tall buildings, leading to the installation of the notorious "suicide nets."

The problems in the production line of digital technology actually begin even earlier, with tales of misery coming from the Democratic Republic of the Congo and Indonesia—places where materials used in hardware manufacturing, such as rare earths and tin, are found. Men, women and children mine these commodities in conditions of slavery, often in the context of conflict. Planned obsolescence in many hardware products means that both the sacrifices made by these workers and the negative impacts on the environment are often tragically redundant.

The development of digital technology has exacerbated some of the worst excesses of the labor market under capitalism: more people receive less pay, with little job security, for unfulfilling work that is both destructive and unnecessary. But labor still remains central to the economy; both the making and the running of technology keep society going.

The traditional Marxist understanding is that working people are essential to the production of value under capitalism. Human labor creates value in a way that all the machinery in the world alone cannot: without workers carrying out the relevant tasks, intellectual or physical, there would be no production. Labor is therefore a critically important source of value. It is also relatively abundant and possessed of its own agency. This theory is fundamental to understanding why the profit motive is exploitative, because the worker is paid only for her time and not for the value she actually produces. "Labor is entitled to all it creates!" as the slogan of the labor movement goes. But this observation is also critical for understanding the centrality of workers—that without them the system of production stops, and that an abundance of labor, or surplus of workers, acts as a disciplinary force. To function, the economy needs people to work, but people are also frightened of losing their jobs, which is almost always their sole means of survival. All this means that labor is a source of oppression under capitalism but also a potential source of power.

The aristocracy represented the ossified capriciousness of wealth distribution that Paine confronted in pre-revolutionary America; Engels denounced the horrors imposed on workers by industrialists in nineteenth-century England. The leaders of technology capitalism fall into a similar category today. Their much-lauded intellect and innovations are directed toward inflating and jealously guarding the wealth they accumulate and helping their fellow captains of industry to do the same. They are indifferent to the miseries they inflict on countless people; they see themselves as noble individuals pursuing their divine right to make money. Yet this is only possible through the exploitation of labor. Which also means that organized labor still holds the power to transform society.

The digital age has created what law professor James Besson has labelled "today's great paradox." He outlines how digital technology has transformed creativity, communications, entertainment and work practices, but one key aspect of our lives that has languished under the influence of new technology is our pay. Contra the previous century, in which technological development overall improved the living standards of working people, it seems that in the last few decades technology has contributed to "a loss of jobs and pay for many ordinary workers, while the pay of top earners has grown dramatically."

Part of the reason for this is that the automation of production has a significant impact on job opportunities for low-skilled workers. For example, automation of the haulage and delivery industry is likely to eliminate most driving jobs in the next decade. As the *Los Angeles Times* observed, "at risk is one of the most common jobs in many states, and one of the last remaining careers that offer middle-class pay to those without a college degree." This kind of warning might seem dramatic—humans are adaptable and have always found ways to redeploy their skills as the economy changes. But the anxiety around these discussions is nonetheless justified. It should be a good thing that the automation of driving will result in safer roads and more efficient logistics and transportation. At the same time, this development promises a world of misery for poor and working-class people, particularly those without qualifications, who will have even fewer decent job opportunities. Commentators often refer to this phenomenon in terms of the "robots eating all the jobs."

But the story is not straightforward. Marc Andreesson, for example, takes pride in debunking this prophecy, with the optimism of a venture capitalist that a truck driver might be less inclined to share. He denies the premise that there is a fixed amount of work to be done (something he spuriously labels as "Luddite"). For him, there will always be more markets—more human needs and wants—to capitalize on. Entrepreneurs have a happy knack of finding new ways to make money, capture more market share, and monetize what has previously been unexploited, off-limits or considered unnecessary. As the door of one industry starts to close, windows to the industries of tomorrow are already opening. The future of technology capitalism is thus accumulative: for the system to continue and grow, entrepreneurs will attempt to commodify more of our lives. They will increasingly try to apply the profit motive to facets of human society that function on the basis of communal values, as we have seen in the improperly labeled "sharing economy." Or they will try to make money by generating useless wants—the over-servicing of the class of people who can afford it. The implication is that anyone left behind in this process exhibits a lack of the entrepreneurial zeal that technology capitalism lionizes above all other human faculties.

The survival of capitalism is also dependent on using technology to extract more from workers wherever possible. Companies are investing in technology that allows them to reduce their labor requirements through data-driven scheduling, and the costs of such arrangements fall to workers to bear. The increasing adoption of this kind of technology is responsible in part for the rise in casualization of labor over the last several decades. Nearly 60 percent of American workers (around 80 million people) are paid by the hour, and nearly half of these are subject to just-in-time scheduling, with no certainty about hours or start times. When we are at work, we are watched more closely than ever. We carry devices connected to the cloud, for example, which monitor every task and the time it takes to perform. The much-derided consumer version of Google Glass has been reincarnated as a workplace management tool, to better track every task assigned to a worker. Amazon is patenting a wristband to squeeze more out of its workers by guiding their movements so they can pack items more quickly. Employers are making the most of technology for the purposes of optimizing how they use and monitor labor.

Automation is creating a world where lower-skilled jobs are much more tightly managed, and higher-paid jobs require rarefied and increasingly complex skill sets. Though it does not necessarily mean a world with less work, the nature and terms of work available through the labor market are changing. The labor market is increasingly characterized by numerous manual, precarious jobs and fewer better-paid jobs that demand cognitive thinking and entrepreneurialism. This is the future that twenty-first-century capitalism presents to us.

Estimates vary, but current figures suggest that globally almost half of all activities people are paid to do at present have the potential to be automated. Very few occupations will completely disappear, but six in ten have at least 30 percent of constituent activities that could be automated. This might mean many things, from requiring a worker to use a computer on the job, to full-scale replacement of the worker. But it will mean that many people will be under- or unemployed, creating a surplus of labor. Statistically, under- and unemployment is growing, and the jobs that do exist pay less and work people harder. If this surplus of labor is a consequence of automation, its existence has always served an important purpose for capitalism. A group of people without work, snapping at the heels of those with unskilled jobs, can be used to corner the working class into submission—depressing wages and creating competition for jobs. The existence of surplus labor means people are forced to accept increasingly precarious working conditions or risk being replaced. There is a socially imposed level of unemployment that companies and governments are prepared to accept, because it acts as a disciplinary force on those who do have jobs.

But it is also important to note that investment in automation is not inevitable or linear. Improvements in efficiency and the elimination of jobs via automation tends to be associated with recessions. For example, the 2008 recession not only saw workers replaced by technology, it also imposed elevated requirements on new hires, who needed IT expertise to work with their new electronic "colleagues." Traditionally low-skill jobs were eliminated and replaced by posts that required more education. This has created what has been called "jobless recoveries," where recessions are replaced by economic growth but with only sluggish growth in job creation. The fallout of this trend, magnified during times of systemic economic stress, affects the more vulnerable sectors of the population. This represents an enormous amount of human suffering; it also reveals the slothful nature of the capitalist class. Rather than rushing to take advantage of new, cost-cutting technologies (the orthodox expectation of how they might respond to market forces), the capitalist class is recalcitrant, only prepared to invest in automation when recessions make the exploitation easier and necessary to justify. They are a class of people who are both torpid and grasping.

Technology capitalism is creating a future of increased commodification and exploitation, which should not be confused with efficiency and progress. The process of automating work, when carried out under capitalism, results increasingly in disparity in income and greater pressure on people to work harder and longer. It also produces a growing class of workers with fewer rights and less security, sometimes called the "precariat," who survive off the gig economy, doing piecemeal work to scrape together a living with no protection from economic change and no long-term security. This system relies on the surplus of labor as an effective deterrent to labor and political radicalism. Many people, including some on the left, like to talk about the centrality of work to our sense of purpose and dignity. David Frayne, in his insightful and intelligent dissection of the modern concept of work, critiques this thesis. Work is in principle how we acquire income and a sense of identity, make a contribution, find community; yet all too often, he observes, "work has also become an extremely unreliable source of these things."

Automation under our current system—where retraining is expensive and the welfare net is minimal—wastes the potential of both workers and technology. It creates the paradox of both increased productivity and impoverishment. Outsourcing work to machines, particularly if the task is boring or dangerous, is a highly worthwhile goal, but it is not one that we are moving toward under the current system. And those who will suffer under successive waves of automation will often be the most vulnerable layers of society—the worst-paid, with the fewest skills and the fewest resources to fall back on in times of crisis. Nick Srnicek and Alex

Williams emphasize the importance of making full automation "a *political demand*, rather than assuming it will come about from *economic necessity*." Technology does not guarantee greater wealth for a tiny minority, nor does it promise a world without work. Its impact will depend on policy choices and who fights for them. But to be humane and just, automation must be part of an agenda in which its benefits are more equally distributed.

What kind of political program would we need to accelerate automation and allow the benefits of this development to be shared better? One of the central tenets must be to work less: to reduce the time devoted to waged labor but without a reduction in wages. We need to reverse the idea that efficiencies gained via automation can be the pretext for greater exploitation by employers.

The idea of a future with less work is not new. Perhaps most famously, John Maynard Keynes, writing in 1930, predicted that the accumulation of capital and the development of technology were leading to a whole new kind of existence for human beings: "For the first time since his creation man will be faced with his real, his permanent problem—how to use his freedom from pressing economic cares, how to occupy the leisure, which science and compound interest will have won for him, to live wisely and agreeably and well."

Keynes believed we were on the brink of creating a society that could meet the subsistence needs of humans—or a society in which human beings had the resources to feed, clothe and care for everyone. The outcome would be a transformation of productive work into something that was minimally necessary but no longer dominant in human existence. Keynes predicted a three-hour workday, or a fifteen-hour week, to spread as thinly as possible the collective task of creating a society of abundance. He regarded the "technological unemployment" already resulting from industrialization and technological development not as something to be lamented but rather as representative of the growing pains of a society moving toward a future of less drudgery and greater output.

Karl Marx, writing even earlier than Keynes, was similarly ambitious about the future of productive work: he imagined the possibility of a workday that lasted six hours, rather than the twelve that were the norm in his time. The consequence of large-scale industrial machinery, Marx imagined, would be the transformation of the role of human labor to the point where productive work became simply "conscious linkages" between the "mechanical and intellectual organs" of automated machinery. For Marx, the outcome of this would be

> the free development of individualities, and hence not the reduction of necessary labor time so as to posit surplus labor, but rather the general reduction of the necessary labor of society to a minimum, which then corresponds to the artistic, scientific etc. development of the individuals in the time set free, and with the means created, for all of them.

The emergence of surplus labor represented a problem for the laboring classes, but it also symbolized the possibility of liberation. Marx pictured a society in which "nobody has one exclusive sphere of activity, but each can become accomplished in any branch he wishes." In other words, a society where it is possible for everyone "to hunt in the morning, fish in the afternoon, rear cattle in the evening, criticize after dinner ... without ever becoming hunter, fisherman, herdsman or critic." Productive work was something that would be increasingly outsourced to machines, with people still supervising production but much more free to pursue a range of interests at leisure.

These writers were representative of a larger, more widespread sense that the future would incorporate less waged labor. Like the technological utopians we met in chapter 5, many people of past generations found the likelihood of a world with less work but more production, sufficient to meet social needs, entirely plausible. To revolutionaries like Marx, to those who like Keynes preferred a regulated form of capitalism, to the many millions who joined the eight-hour-day movement and countless others, less work seemed self-evidently like a progressive step toward a better society.

We are not inching closer to this reality; we are drifting further away. More people are working longer hours, in more jobs. Automation under capitalism is not oriented toward living "wisely and agreeably and well." Instead it has created a deep, abiding moral order that binds merit to hard work. This is a culture that starts at the very top—the ruling elite of the twenty-first century are no longer slovenly landed gentry or lazy heirs to vast fortunes. They are intensely hardworking, and they believe that gives them an entitlement to wealth. This attitude was chillingly exemplified in an anonymous Wall Street trader's letter to Occupy in 2011:

> We are Wall Street … We get up at 5am & work until 10pm or later. We're used to not getting up to pee when we have a position. We don't take an hour or more for a lunch break. We don't demand a union. We don't retire at 50 with a pension. We eat what we kill, and when the only thing left to eat is on your dinner plates, we'll eat that.

We also find it in the start-up selling "Soylent," a nutritionally balanced meal replacement that allows busy Silicon Valley types to free up time that might otherwise be wasted cooking. The body-optimizing and time-hacking ethos of Soylent sluggers shares something with the trader's invective: a smug belief in the cultural and moral connection between productivity and self-worth. Keynes foresaw this problem: "We have been trained too long to strive and not to enjoy," he wrote. The exclusive valorization of a meritocratic work ethic promises misery, overproduction, inequality and the marginalization of anyone who is unable to participate in this competition for exploitation. It is a mind-set we need to resist.

Working people have the capacity to do this. As an essential element of capitalist production, they also hold the power to shut it down. It is time to revive the international demand for shorter work hours, for the reduction and equitable redistribution of waged work. Our goal should be a six-hour day, leading to a five-hour day; a three-day weekend en route to a three-day week. We need to reclaim time from the destructive and exploitative work ethic that dominates current conversations about labor.

Frayne calls for a return to the *politics of time*: "a concerted, open-minded discussion about the quantity and distribution of working time in society, with a view to allowing everybody more freedom for their own autonomous self-development." This also happened to be a key motivation of the eight-hour-day movement: to minimize unemployment through the equitable distribution of waged work. At the unveiling of a monument in 1903 to celebrate the historic achievement at Melbourne University half a century earlier, the labor organizer Tom Mann declared that the aim of the movement was to abolish poverty and do away with the unemployed, and the only means of accomplishing that was to have a shorter working day. (He called for a six-hour day, and reports indicate that the crowd joyfully concurred.) Relaunched today, such a movement would also create employment opportunities for the people who are currently considered surplus to requirements by the capitalist class, and thus undercut the capacity to use those spare workers to keep the others in line. In other words, as Frayne puts it, "everyone should work less so that everyone can work."

Organized labor has traditionally demanded both bread and roses—the right to decent pay but also conditions of dignity and beauty. As well as reducing the hours devoted to work, the stonemasons in Melbourne were motivated by the desire to transform work into something that was carried out under reasonable conditions that allowed workers to flourish personally. A shorter workday was necessary to deal with the harsh climate of Australia but also reportedly "to give the worker time to read and study, and to progress in knowledge and virtue"; to make them better people, inclined toward "self-respect, and respect for other[s]" and "domestic virtues and good citizenship." The campaign for an eight-hour day was "a public necessity," as another writer put it, "as a means of giving to men a wider interest in life, the possibility of greater culture, and the surety of education commensurate with the problems now forced upon our democracy for solution."

Taking inspiration from this framing, we can demand an end to technologies of workplace organization that squeeze more from workers at the expense of society, such as data-driven scheduling and surveillance technologies. This kind of technology creates a harsh climate of uncertain and irregular hours of frenetic work, with families and communities bearing the consequences. Work needs to be a place where people enjoy conditions of respect rather than discipline. We should ensure that improvements in productivity that have come about through automation are enjoyed by all those who work and contribute to production, not just converted into profit for their employers.

We can also use this discussion to campaign for time for other pursuits, away from waged work. The law should drastically increase the costs of overtime and impose a prohibition on work beyond a weekly cap of hours. This has a specific relevance for the digital age: France recently instituted legal protection of "the right to disconnect," meaning that workers have no obligation to send or respond to emails outside of working hours. This has taken the form of server shutdowns overnight and automatic deletion of emails sent outside of business hours. A similar law is now being mooted in New York City. In the Netherlands, workers have the right to demand reduced hours and cannot be penalized for exercising this right. It is possible to imagine other legislative proposals that incrementally dismantle the culture of long hours and allow working people to learn new things, contribute to public policy discussions, and become greater contributors in their communities and homes.

The demand to reduce working hours and redistribute socially necessary labor requires us to reorient: from understanding value in terms of money to measuring value in time, well-being and sustainability of the planet. Reducing the number of working days has been credibly shown to reduce climate emissions, by cutting the emissions associated with commuting and the energy costs of running workplaces. The State of Utah experimented with this by reducing the days worked by state employees for a number of years, and researchers documented the reduction in greenhouse gas emissions.

Public discussions around the politics of time would also help us find ways to minimize pointless work. It would force us to consider what kind of work might be necessary for a functioning society and what kind of jobs are actually "bullshit," as the anthropologist David Graeber describes them. It would mean our collective decisions about production were actually responsive to the climate we find ourselves in and our needs as a society. The market is evidently a terrible decision-maker in this respect, because, as Graeber puts it: "In our society, there seems a general rule that, the more obviously one's work benefits other people, the less one is likely to be paid for it." It would be a world with more nurses and teachers, fewer corporate lawyers and stock traders. Fewer speculative entrepreneurs and more people working together to address urgent, concrete problems like climate change at a global and local level.

Just as the original movement for a shorter working day took off in the decades after the industrial revolution, so too can this claim be made legitimately today, in the wake of technological developments that elevate our collective productive capacity. It would be an example of the living legacy of what began with Paine, was demanded by Marx and was imagined by Keynes. It would be the continuation of a tradition that attracted millions of followers in the form of the eight-hour-day movement. It will take a new system of production—where we encourage automation and reduce working hours without loss in pay—to allow the ideal society with the minimum of socially necessary work to come into being. We need to borrow from our forebears and adapt their tactics to our present situation. We need a movement that calls on digital technology to allow one to become a hunter, fisherman, herdsman and critic all in one day. We need a society in which human life is not judged by its productive output or income or wealth, one in which technology is directed toward creating space for the multiplicity of the self to blossom.

The potential of automation to create a world with less waged work is exciting, even if we can only see glimpses of it today. Digital technology has the capacity to create a world where production is more efficiently carried out by machines, but an essential part of achieving this goal is to give more power to workers in carrying out their jobs. If workers are given the chance to engage meaningfully with their work, to learn skills and test them in practice, they can feed that knowledge back into productive processes in ways that are highly effective. The transformation of work via automation does not necessarily mean a fixed future of robots eating jobs, but it does create an opportunity for robots to eat the rich.

"Technology implementation is challenging," Besson writes, "because large numbers of workers need to acquire new skills and technical knowledge." Technological development in these contexts is best thought of not just as bright ideas, such as the steam engine or the microchip, but as a combination of those bright ideas and their implementation or maintenance. Education in this context is not simply a matter of degrees or qualifications but rather engagement with technology that builds skills and knowledge. Put another way: people working with new inventions and technological developments often contribute significantly to the improvement of these inventions over time. By working with technology collectively, we are more likely to improve that technology and get the best out of what innovations have emerged.

Worker cooperatives offer a format for production that can efficiently adapt to new technology. They come in many different forms, but, in general, worker cooperatives are enterprises in which workers own the bulk of the firm collectively and management decisions are made with the democratic participation of the worker-owners. Such organizations exist all over the world, particularly in North America, Latin America and Europe, in a range of industries and in varying sizes. A good example came out of the attempted closure in 2008 of Republic Windows and Doors in Chicago. The business was still profitable, but the owners wanted to shut the company down and set up a similar business using temporary labor. The workers fought back: resisting the dismantling of the business, they eventually gathered together the capital to purchase the company themselves. There are now twenty-three worker-owners running the business successfully. They renamed themselves New Era Windows Cooperative.

The driving force behind these enterprises is active participation in management decisions by workers. This results in not just a broader set of ideas about how to organize labor; critically, it generates knowledge-sharing about work—something that is highly valued but difficult to cultivate in more orthodox business structures. Structuring organizations in this way creates the strongest incentive among workers to be creative, take responsibility, avoid unnecessary conflict, and monitor and improve performance. In other words, to educate themselves and each other on how best to work with technology. Worker cooperatives are places where workers can play with new skills. They can learn, practice and experiment with new ways of doing things. They can share knowledge and feedback, confident that they will also share in the benefits of any higher productivity that may come about as a result. "If we make a mistake, we talk to each other and we find a solution," explains Armando Robles, a worker-owner of New Era Windows. "We try to do the best for everyone. We work harder because we're working for ourselves. But it's more enjoyable."

Again, here we see the seeds of the future being planted in the present. So much of what is often called the sharing or gig economy—enterprises like Uber, Airbnb, Zipcar, Task Rabbit—are well suited to being modeled as worker cooperatives rather than profit-making companies. These workers themselves understand the possibilities of organizing themselves. Knowledge-sharing networks exist for Uber drivers, for example, that help them navigate the complexities of their working environment and help researchers to document the challenges of this form of employment and make them visible to the public. For obvious reasons, often companies have tried to ban or marginalize these initiatives. Encouraging them and providing them with the protection of law is an essential demand for a modern labor movement. There have also been some experiments in organizing these kinds of platforms directly as cooperatives. They have produced mixed results, but the prospect remains promising, particularly in models that give the most control in management decision-making to workers.

The gig economy has often used digital technology to scale up specific social values, including the value of self-directed work, the efficient use of limited resources via sharing, the importance of trust and an interest in building a community. If we imagine the possibility of re-socializing these models of work, through giving control back to workers, we glimpse a promising vision of the future of work.

Unfortunately, under our current system it is well-documented that the gig economy is full of highly exploitative companies—exploiting both the people who rely on them for a livelihood and the communities in which they operate. When geared to making money, companies in the sharing economy end up risking the destruction of the very values that made them so successful in the first place, such as openness, a sense of community and goodwill, and a willingness to share. The problem is when companies capitalize on these values to pursue profit rather than distribute the benefit to others. They operate according to the rules of the marketplace rather than rules set by a democratic process. But this outcome is only possible because of a certain approach to history that has permeated the age of digital technology. As Tom Slee points out: "The Sharing Economy's complete neglect of the history of collaborative and co-operative movements is one of the reasons it has been so easy for business to co-opt."

We are already moving into a world where work can be be flexible, self-directed, prioritize the sustainable use of resources, and be based on trust and community-building. It is possible to imagine how gig enterprises could be transformed into worker cooperatives to ensure that the benefits of working in this way accrue not just to company owners but also to the people who are actually doing the work. To do so might require certain interim policies, such as tax breaks for cooperatives or finding ways to facilitate the transition to worker ownership when a company becomes insolvent. It could also take the form of unions demanding a form of ownership, even control of members' workplaces. When workers have a greater opportunity to participate in decisions about how things are done, it is easy to see how the worst kinds of work will gradually be prioritized for automation. "Bullshit jobs," like flunkies for management or fixers of nonexistent problems, would be deprioritized, even rendered unnecessary once there was greater feedback between people making strategic decisions and people doing the work. The necessary skills for adapting to new technologies would be openly shared. This model takes the advantages of digital technology—its capacity to scale up a sense of trust and community—and distributes them far more broadly than is the case at present.

The deep sense of solidarity and collaborative nature of worker cooperatives creates conditions for a bold and active vanguard of working people able to generate demands for automation and redistribution. Traditional unions are still critical, but we also need to think more expansively about organizing in the age of automation. Many of the companies that are growing exponentially in the digital age are well suited to these kinds of structures. Companies that rely on the sharing economy offer a glimpse of how the future might look, but not for the reasons that their CEOs like to think.

A more democratic approach to automation will prioritize politics that reduce exploitation and inequality, and look outside of market forces to make decisions about how we produce things. But even in a world with less waged work, there will be inequality, as some kinds of labor will be valued in productive terms more than others. We need an agenda for a society that ensures that people both contribute fairly (working in conditions of fairness, not exploitation) and are adequately provided for (with resources redistributed according to needs, not market incentives). Reducing hours is an important part of achieving the former. In respect to the latter, Thomas Paine might argue that it is high time that we demand that a ground rent be paid.

There is a divide among the left over the form that this ground rent should take—that is, what kind of policies are best suited to redistributing the benefits of automated production. Some authors, like Nick Srnicek and Alex Williams, argue for accelerating the process toward full automation as part of a modern left vision of a post-scarcity economy. Others, like Adam Greenfield, are more skeptical about the likelihood of such a prospect. As Greenfield points out, some of the most radical sections of the working class constitute the very parts of the economy that will be automated the quickest. If we do aim for a world with less waged labor, what kind of redistributive mechanisms should we implement to ensure that this is empowering rather than immiserating? What does a modern equivalent of a ground rent look like for the digital future?

I would argue there are three things we need to consider: how to ensure people are paid properly, how to organize productive work responsibly in ways that recognize the importance of other kinds of work, and how to ensure that people's basic needs are met. There is a range of different ideas and proposals to help us think about how these goals can be achieved.

The basic income (often called a universal basic income) finds its philosophical roots in the concept of a "groundrent" floated by Paine over two centuries ago. Writing in 1920, Dennis Milner was arguably the first person to formulate an explicit argument in favor of a basic income, an argument that took the form of a pamphlet circulated among the British Labour Party. Charity was not the answer to poverty, it said; higher production would only come about when certain conditions were met. Milner argued for the twin needs of "food and freedom": citizens need food and shelter as well as freedom to control their lives. These needs could be fulfilled through regular payment of a sum of money to all citizens, without qualification.

Nearly a century later, similar proposals have taken hold in a number of countries. A basic income has been or is being considered by major social democratic parties in the UK, where it enjoys popular support, and Canada. A proposal was considered in Switzerland. An experimental program in Finland, introduced with significant fanfare, was subsequently abandoned. (That controversial decision struck many commentators as premature and appeared to be justified on the basis of poor design.) Meanwhile, a form of it has existed in Brazil for decades.

But the concept also has right-wing proponents. Milton Friedman was a supporter, on the basis that it would reduce the role of government. More recently, the elite of technology capitalism, including Mark Zuckerberg, Marc Andreessen and venture capitalist Sam Altman, have backed the policy. Libertarians have proposed doing away with the welfare state entirely, to be replaced by a single, flat-rate universal payment. From this perspective, a basic income serves to safeguard the long-term prospects of capitalism and accelerate speculative entrepreneurialism. It serves as a facilitator of free-market functionality rather than a limitation on market extremes. In this way, it has the potential to paper over the structural features of technology capitalism that exacerbate inequality. There is no brake on the expansion of consumerism, no slant toward sustainability in a context where market forces dominate decision-making about public resources and collective problems. The people who will pay the price are those who make up the class of surplus labor—those whom the market renders expendable, who struggle to find work in this fragmented economy. If a basic income becomes the excuse to replace universal social programs, like education, health and welfare, as part of liberating markets, it will herald a form of death-cult capitalism.

This approach is emblematic of how Silicon Valley's understanding of poverty is seen as a design problem to be tinkered with and solved by structuring redistribution in ways that ignore the underlying problem. These thinkers are enamored with the free market, ignoring its relentless drive to colonize every aspect of human existence, glossing over its incapacity to effectively manage common resources and its structural tendency to externalize a range of economic, social and environmental costs. The elite of technology capitalism want to use a basic income to soften the edges of the labor market in a time of technological change, so that they can continue to do what they do best: exploit people for profit.

Is, then, the basic income a radical redistributive proposal or is it a way for capitalists to preserve their power? The answer will depend on who demands it and how. It could be a technocratic lifeline to subsidize rentier capitalists, or it could give everyone a substantively equal share in the productive output of a sustainable society. Either way, it is a debate that we appear to be having. Rather than outright reject it, the left needs to respond to the prospect of a basic income as a starting point for further demands for a better, more sophisticated approach to redistribution of wealth. Failure to do so will allow the contours of the debate to be defined by the elite, who are increasingly discussing it on their own terms.

In this context, it is critical that we understand the benefits of a basic income. The first concerns the role of the judgmental and time-consuming bureaucracy around welfare, the effect of which is to make government assistance difficult and humiliating to claim. The introduction of a basic income would be a marked improvement on current arrangements.

A basic income would also involve a kind of welfare that has repeatedly proven to be effective: giving poor people cash. Cash payments are highly efficient, particularly as there are few implementation costs, and cash has multiplier effects that the delivery of goods and services does not create. Allowing people agency over their welfare provision and expenditure has the potential to be empowering.

Perhaps most importantly, the concept of a basic income also gives legitimacy to a great deal of work that has traditionally been dismissed as nonproductive. Implemented as a universal proposition, a basic income is not attached to waged employment, as are tax credits or statutory work entitlements. Reproductive labor in the home, such as care work and housework, has traditionally been invisible to the economy. A basic income offers a form of payment and creates the potential for this labor to be valued financially for the first time in a meaningful way. It also has the potential to liberate the many people (mainly women) whom this kind of duty often excludes from other earning opportunities and confines to a life of abuse, control and isolation. A similar argument can be made for individual endeavors that benefit the community as a whole, including making art, volunteering, political organizing, and public participation in policy-making. Public participation in decision-making could be elevated in all sorts of new and interesting ways if the vast majority of people were relieved of their dependency on waged work for financial security.

In other words, a basic income has the potential to undercut the Wall Street mantra about growth and jobs, productivity and individual worth. It is predicated on an understanding of productive work not as something indispensable to our sense of value but as something that should be limited to what is socially necessary and increasingly just a small part of what makes us human.

Alongside this, we need to consider how productive, waged work could be more democratically organized to meet the needs of society rather than individual companies. To this end, one commonly suggested alternative to a basic income is a job guarantee. The idea is that the government offers a job to anyone who wants one and is able to work, in exchange for a minimum wage. Jobs could be created around infrastructure projects, for example, or care work. Government spending on this enlarged public sector would act like a kind of Keynesian expenditure, to stimulate the economy and buffer the population against the volatility of the private labor market. Modeling suggests that this would be more cost-effective than a basic income (often critiqued for being too expensive) and avoid many of the inflationary perils that might accompany basic income proposals. It also could be used to jump-start sections of the economy that are politically important, like green energy, carbon reduction and infrastructure. A job guarantee could help us collectively decide what kind of work is most urgent and necessary and to prioritize that through democratically accountable representatives.

But a job guarantee, when compared to a basic income, has its own drawbacks. It reinforces the connection between productivity and human value. There is a real risk of more bullshit jobs being created under a job guarantee policy, and, if poorly implemented, it could deny people the time to engage in other pursuits, whether care responsibilities or artistic endeavors. It has the potential to leave behind people who cannot work, perhaps due to disability or other health concerns. There is something unavoidably didactic and potentially disciplinary about the concept. Frayne urges a focus on policies that do not "seek to enrol people in some pre-planned utopian scheme, but to gradually free them from prescribed roles, furnishing them with the time to become politically active citizens." We should be wary of the prescriptive approaches to policy-making that underpin a job guarantee.

Still, the idea of a job guarantee is a useful one to consider. It hints at something important: the centralized organization of socially necessary work. The universality of the idea has other appealing qualities. It could be used as the basis for a right to certain kinds of universal leave, such as parental leave, sick leave and long service leave, transferable between employers. Anyone who works and meets the minimum requirements of a job under the guarantee could be paid these entitlements out of a publicly managed fund collected from employers, for example. It creates the prospect of collective decisions about production that are not based on profit but on human need, environmental sustainability and dignity in work.

To be liberating, rather than enforcing a paradigm of work-for-welfare, a job guarantee must allow for flexible and expansive definitions of work that satisfy the requirements for inclusion in the scheme. The working hours should be fewer than the current average working week. It should aim to organize socially necessary work and distribute it among the population able to carry it out. It might start to look like the world Bertrand Russell imagined when he argued for a four-hour day, in which the work "will be enough to make leisure delightful, but not enough to produce exhaustion." In short, it ought to look a little more like a basic income for socially necessary work.

As part of this mix, we also need to consider the role of service provision, since neither a basic income nor a job guarantee necessarily provide for this. This idea has been called "universal basic services," that is, the provision of health, education, housing, public transport, connectivity and personal care services to everyone, free at the point of service. Rather than focusing on a monetary or financial method of redistribution, it represents a shift to a service-orientated model designed to meet the needs of people more directly. This approach to welfare provision acts as a counterpoint to the idea that a basic income replaces these systems of welfare. It also ensures that anyone who might not be able to perform socially necessary work is cared for properly. The stigma associated with welfare services is reduced, as everyone pays into and receives the benefits of social insurance.

Ultimately, we need a combination of these programs. We need the liberty offered by a basic income, the sustainability promised by the organization of a job guarantee, and the protection of dignity offered by centrally planned essential services. It is like a New Deal for the age of automation, a ground rent for the digital revolution, in which the benefits of accelerated productive capacity are shared among everyone. From each according to their ability, to each according to their need—a twenty-first-century vision of socialism. "We have it in our power to begin the world over again," wrote Thomas Paine in an appendix to *Common Sense*, just before one of the most revolutionary periods in human history. We have a similar opportunity today.

9

We Need Digital Self-Determination, Not Just Privacy

Frantz Fanon Theorizes Freedom

"O my body!" begs Frantz Fanon, in the final line of his ground-breaking book, *Black Skin, White Masks*. "Make of me always a man who questions!"

Fanon was a thinker who never shied away from profound and abstruse questions about power, race and identity. Born in Martinique in the Antilles in 1925, when it was a French colony, he served in World War Two with the Allies. He had lived a middle-class life in the Caribbean, but his military service was a rude awakening: his blackness trumped his French citizenship among his fellow soldiers, and he found his identity predetermined by his race. After the war, he traveled to France and eventually trained as a psychiatrist in Lyon. In 1953, Fanon was sent to work in Algeria, and within a year the War of Independence broke out. Exposed to accounts of torture by Algerians and already sympathetic to their cause, he sided against the colonial French. He left his job, returning to France briefly before heading to Tunisia to work full-time for the provisional government of the Algerian Republic, committed to the anticolonial revolution. His posting eventually took him to Ghana, which was then the social and political center of the African movement for unity. All the while, he was theorizing and writing about what he thought and saw.

These firsthand experiences, together with his training in the theory and practice of psychoanalysis and his impressive skills as a writer, meant that Fanon produced riveting and original work with lasting impact. In an ambitious yet accessible style, he discussed far-reaching ideas in tight and compelling prose. His writings grappled with the nature of colonialism and racism and what kind of future was possible for people who have lived through these systems of oppression. He identified anti-black racism as a form of skewed rationality, a systemic problem, rather than an essential feature of human society. He created a language for talking about these ideas, and he theorized about their causes.

Fanon was committed to the idea of self-determination. Even in a world in which human beings are categorized, surveilled and discriminated against, it is possible for us to carve out space for our own identity and shape our destiny. Though his analysis was informed by his experience of being black in a racist society, Fanon's oeuvre was never defeatist. In diagnosing the complex and deep-rooted nature of racism, he did not advocate a defensive retreat into a fixed sense of identity or into racial nationalism. His ultimate vision was emancipatory: he always questioned not only why racism existed but also how it could be transcended. The loss of dignity was reversible for Fanon, through struggle.

He fell ill with leukemia shortly after travelling to Ghana; and after some travel for treatment, he died in 1961. One of his other key works, *The Wretched of the Earth* (the title taken from the first line of the revolutionary anthem "The Internationale") was published posthumously. His death marked the end of a six-year spell on France's enemy expatriate list, and a lengthy stint on the assassination lists of several French proto-fascist groups.

Fanon's explanation of how colonialism works—how both the colonizer and the colonized take on roles and practices that make colonialism seem inevitable and unassailable—has relevance well beyond his time. He argued that liberation was possible, even in the context of significant oppression, and this argument continues to find applicability after his death. Indeed, I would argue that Fanon's theoretical work can be repurposed in order to make sense of digital life in the twenty-first century. The questions Fanon grappled with continue to present themselves, namely: How do we create ourselves in a world in which our identity is predetermined for us? How can we find freedom in a digital environment where the history of our sense of self—our abstract identity—has been written by others? Those who hold power over our digital lives, from private industry to government, do so in a way that is almost hegemonic. As we have seen in earlier chapters, the eyes of technology capitalism watch everything we do, and they share their insights with the forever-snooping government. Our behavior is tracked and predicted, we are categorized and judged. How do we define ourselves outside a paradigm—which categorizes us as consumers and objects of surveillance—that is so powerful that it is virtually inescapable?

To be sure, the experience of having our minds colonized is not comparable to the iniquity, violence and horror of actual colonization or to the pain of living in a deeply racist society. Relating Fanon to the problems of the digital age reflects an attempt to take his work seriously, to apply his thinking outside its time and place, much as we do with other thinkers, like Marx or Paine or Freud. Moreover, I would argue that there is something dehumanizing about how technology capitalism has presided over discrimination and impoverishment, using potentially transformative technology to usher in a new Gilded Age. There is violence in how the government uses technology to incarcerate and kill while claiming its practices and weaponry are more precise and neutral than ever. There is something deeply disgraceful in how our reliance on the network has been weaponized by the surveillance state as part of a cyberwar arms race that puts the poor and vulnerable at risk. These processes may be in their early stages of historical development, but they also deserve to be taken seriously. "Fanon dedicated his life," according to the philosopher Louis Ricardo Gordon, "to breaking free of the weighted expectations of consciousness without freedom." This objective ought to be equally applied to modern digital life.

The complex process of evading government and corporate

surveillance is often framed as a right to privacy. But as discussed in chapter 1, the right to privacy fails to capture the kind of freedom we need in digital life. Privacy is typically conceived as the right to protect one's individual world or home, sealed off from the public, away from view. But isolation is not the same as freedom. Instead we need to find ways to create a collective social and political life that prioritizes autonomy and is not subject to scrutiny by those who seek to profit from it.

We need to wrest digital self-determination from surveillance capitalism and the state. Digital self-determination will involve making use of the available technical tools to communicate freely. But it will also require autonomy, away from manipulation by the vast industry of data miners and advertisers that tries to shape our identity around consumption. We need to change the legal structures governing data and its use and design information infrastructure in ways that favor decentralization. Digital self-determination must also be about designing online spaces and devices that are welcoming and give us control over our participation rather than creating a dynamic of addiction. It is about the freedom to question, to freely determine our sense of self, both in digital environments and outside of them.

Fanon wrote about seeing himself, a black man, in a world ordered by white supremacy. He explained how his identity was not his own, how the white man had "woven me out of a thousand details, anecdotes, stories." He viewed his own self through the prism of white supremacy:

> I was responsible at the same time for my body, for my race, for my ancestors. I subjected myself to an objective examination, I discovered my blackness, my ethnic characteristics; and I was battered down by tom-toms, cannibalism, intellectual deficiency, fetishism, racial defects, slave-ships, and above all else, above all: "sho good eatin'."

He was, in short, a collection of stereotypes and assumptions, drawn from past experiences of that same system of supremacy. The experience of being black was, by virtue of being black, not an experience. Racist logic rendered it nonexistent—defined by others, the subjective experience of the black person could not be bridged or translated into the real world. There was no agency in how his identity was determined, no way to escape the judgments about him, no glimmer of autonomy. His identity was "*fixed*": "I am overdetermined from the exterior. I am not the slave of the 'idea' that others have of me but of my appearance." This is the theoretical basis of colonialism, and it is how the idea of race is socially constructed.

These ideas also apply to the experience of living in the digital age. How we appear online—our abstract identity—is being generated and fixed by the data mining industry. It is shunting us into reputation silos and being used by the government to make decisions about us. We must urgently find ways to live free from presumptions and categories, ways to limit how the twin powers of surveillance capitalism and the state seek to define our sense of self.

Digital technology already exists that allows us to impose some limits on this power. We can evade some of the invasive practices of surveillance that define our abstract identities. Technologically, we are already able to communicate in secrecy and anonymity: we can use encryption and "The Onion Router" (Tor). Encryption especially has become commonplace, as a standard requirement of mail and web browsing after the revelations of NSA surveillance by Edward Snowden. For better or worse, this is also a function of capitalism: because encryption underpins the operation of the international banking system, its widespread adoption is vital to the economy. Similarly, Tor is (at least in part) a product of the military and its need to communicate anonymously. To put it differently, privacy-enhancing tools often serve multiple purposes, which is why they are improving and getting easier to use. But there is more political mileage to be gained in this space for the left if we put our minds to it.

There is still work to be done in ensuring that this kind of technology is used more widely, to resist the power of companies and governments to define our sense of self. Technologists need to build bridges with everyday people and help popularize these tools, with empathy and humility. People without a technological background need to learn more about what they are up against. The left's job is to build communities and organizations where people can both learn and teach tactics for digital self-defense.

My point here is that, from a *technological* perspective at least, certain kinds of privacy are possible—namely secrecy and anonymity. But to some degree this misses the point: it means very little if people still spend significant time on social media sites or apps and platforms that collect and use data that can undermine the usefulness of these tools. Most people do not face challenges of the kind Edward Snowden did after he leaked NSA documents—very few of us will find ourselves in situations where we are trying to communicate privately while under direct scrutiny by the world's largest intelligence agencies. Most people have to live in the real world, where many everyday transactions require giving up some privacy. This is why many of us, when confronted with the enormity of state surveillance and the monumental effort it takes to personally circumvent it, often give up. Why not give Facebook your data, if all that happens is you get better ads? Why bother with slow and fiddly browsers and complicated passwords if you're a nobody the government is unlikely to care about? What if you cannot afford to learn about various privacy tools that seem designed for only the tech-savvy? For these reasons, people often end up tolerating surveillance as a neutral or relatively benign phenomenon, or as a tool used on individuals who warrant being watched—people who have something to hide. Privacy gets depoliticized. Its supposed importance starts to wear thin.

The kind of privacy that can be protected by these digital tools reflects only a thin slice of the kind of freedom we should be able to enjoy in digital society. We need to reframe how we understand the problem of surveillance at a remove from the purely personal or straightforwardly technical right to privacy. We also need to junk the associated, underlying premise that it is necessary to give up privacy in the interests of security.

With reference to his experiences in Algeria, Fanon wrote about how the "degrading and infantilizing structures" of colonial relations disappeared during the independence struggle. "The Algerian has brought into existence a new, positive, efficient personality, whose richness is provided ... by his certainty that he embodies a decisive moment of the national consciousness." Fanon used this idea metaphorically: the new personality of the Algerian was autonomously self-defined. By opening up the idea of social relationships outside oppressive colonial traditions, the struggle in Algeria created space for a whole new sense of personality to come to the fore.

In Algeria, this took a specifically technological form. Fanon wrote about how prior to the war of independence, radio had "an extremely important negative valence" for Algerians, who understood it to be "a material representation of the colonial configuration." But after the revolution broke out, Algerians began to make their own news, and radio was a critically important way of distributing it cheaply to a population with minimal literacy. Radio represented access not just to news but also "to the only means of entering into communication with the Revolution, of living with it." Radio transformed from a technology of the oppressor into something that allowed Algerians to define their own sense of self, "to become a reverberating element of the vast network of meanings born of the liberating combat."

To reapply this in a modern context, therefore: digital privacy—and its philosophical twin, freedom—involves anonymity, secrecy, *and autonomy*. Autonomy is not just evading surveillance. Autonomy means the freedom to act without being controlled by others or manipulated by covert influences. This kind of freedom is not only jeopardized by spooks and cops. It is also being eroded by the practices of technology capitalism. Our understanding of privacy needs to engage with the imprint we leave on the web collected by companies, categorized and manipulated. Just as the Algerians took control of their sense of self using the technology of radio, so too can we do something similar with digital platforms today.

Bernard Harcourt talks about us all having "digital doppelgangers," or algorithmically matched versions of ourselves. They follow us around digital spaces, reminding us of what we have done and channeling us into a particular future. This process of abstract identification is built on assumptions that are path-dependent, accumulating data mindlessly, beyond our control—and increasingly an inescapable part of the modern experience of being human. People find dates online, they keep in touch with relatives using social media, they visit health sites to research embarrassing conditions, they buy things, they apply for loans using the web and perhaps miss a payment or two. We cannot expect the problem of diminishing privacy to be resolved by social abstinence and daily inconvenience. For privacy to be meaningful, it needs to be about winning back control over our own sense of self—demanding our rights collectively. It needs to drive a stake through the heart of these zombie digital doppelgangers.

A better way to understand what we mean when we talk about privacy, then, is to see it as a right to self-determination. Self-determination is about self-governance, or determining one's own destiny. Its origins as a legal concept stretch back to the American Declaration of Independence, which states that governments derive "their just powers from the consent of the governed." It has always featured as a right of some description in international law, usually in the framework of nationhood and governance of territory. But with the explosion of postcolonial struggles in the latter half of the twentieth century, it gained new meaning—not least in the struggle for Algerian independence that Fanon was involved in. In places like South Africa, Zimbabwe (then Rhodesia), the Democratic Republic of Congo and others, mass social movements struggled for recognition outside the confines of colonial settler states. Later these places often found themselves burdened with postcolonial systems that reproduced familiar hierarchies. The right to self-determination took on a renewed and deeper urgency, raising questions about how to empower people culturally, socially and politically, outside of the European ideals that offered lofty language but had also legitimized colonialism.

Self-determination in this latter sense is the kind of thinking we need to take into the twenty-first century. Rather than seeing it solely as about the right to vote or national governance, it is a right that can find new meaning with the help of digital technology. The ideals of nationhood have (justly enough) less relevance in digital environments. Self-determination is both a collective and individual right, an idea of privacy that is much more expansive and politically oriented. It is about allowing people to communicate, read, organize and come up with better ways of doing things, sharing experiences across borders, without scrutiny or engineering, a kind of cyberpunk internationalism. If we think about the practice of abstract identification discussed in the earlier parts of this book—the practice of weaving us out of a "thousand details, anecdotes, stories," in Fanon's words—we need to think about how we can wrest back control of these processes, to give people the power to define their own identities, a form of personal data sovereignty. We need to transform the technology of surveillance and oppression into a tool of liberation for defining ourselves autonomously.

So what could digital self-determination mean, in practical terms? The first demand must be that public and private actors be held to meaningful standards of disclosure about how information is collected, stored and used. Everyone should have the right to know what is known about them. Everyone should have a right to see the data that go into these processes of identification and in doing so have the power to alter them. Self-determination ought to include the right to meet your digital doppelganger, as a way of understanding ourselves, as a way of regaining control of our identity. Such transparency would mean that companies are better able to be held to legal standards of nondiscrimination, much in the way that banks were the focus of community activism and legislative reform to address redlining in the mid-twentieth century.

We should consider resetting the legal nature of our relationship with companies that collect and hold data from and about us. Law professor Jack M. Balkin argues that we should think of these companies as holding our data in the same way a doctor or lawyer would—that is, by virtue of a relationship of trust. "Certain kinds of information constitute matters of private concern," writes Balkin, "not because of their content, but because of the social relationships that produce them." He argues that we should think of these companies as information fiduciaries, and just as we would not allow our doctor or lawyer to sell information about us to data brokers, the same restrictions should apply to companies. Under this area of law, fiduciaries owe a duty of care and a duty of loyalty, and breaches of these duties are penalized by courts. The kind of information held about us by companies is personal, and potentially damaging if made public; it ought to be subject to similar regulation.

Companies should not be permitted to sell data we give to them to third parties, and we should not be able to contract out of this via terms of service. It would be illegal for a doctor to sell medical information, even if the patient consented. The same ought to apply for our abstract identities. Such a measure could strike at the heart of surveillance capitalism's business model and all the associated technologies of state surveillance that draw on the data collected as a result of it.

Recasting our personal legal relations with the data mining industry is an important first step. But it must be more radical than a mere renegotiation of rights. "The Algerian combatant is not only up in arms against the torturing parachutists," wrote Fanon. "Most of the time he has to face problems of building, of organizing, of inventing the new society that must come into being." It is not enough to address wrongdoing or limit the worst excesses: we have to be more ambitious in our thinking. Digital self-determination also asks us to rethink how we design and build our information systems in aid of a more democratic society. It forces us to question who we allow to be the gatekeepers of our knowledge and how this power could be redistributed. We have to actively build, organize and invent a new digital society.

The systems as they are currently designed are vulnerable. The centralized approach to managing personal data underpins surveillance capitalism. It allows companies to accumulate data and draw conclusions that can be used for segregated marketing. This is not a consumer choice problem: the terms of service are imposed upon us rather than freely entered into.

But there are alternatives. It is possible to design data storage systems that are decentralized and give individuals control over their personal data, including its portability, and the ways in which it is shared with third parties. One example of this idea in action is Solid. Analogous to a post office box, Solid aims to create pods of data. Instead of handing over data to companies wholesale, either directly or by granting permission to access devices, the idea is that users give permission to certain applications to access certain information, and that information can be updated as necessary. The outcome is that an individual can create their own personalized terms of service. These kinds of systems allow data to be stored anywhere—in the cloud or on a hard drive—while also giving the individual full control over it. It is an example of data sovereignty—control over one's personal information, creating the proper basis for meaningful consent. Other examples of projects that center on the interests of users include Freedom Box, a router with open source firmware built in to give users control over and protection for their data, and Diaspora, a decentralized, federated social media platform. Just like how Algerians repurposed radio—from a platform of the colonialist project into a tool for liberation—we too can adapt existing digital infrastructure to decentralize data flows and take back power from companies and governments.

These alternatives aim to "re-decentralize" the web—that is, to return to its original architecture, before its centralization into large platforms. These projects make up "a robust and fertile community of experimenters developing promising software" in this space, which involves "deeply exciting new ideas." To become viable alternatives, these software programs will need to overcome serious hurdles, including proper resourcing and management, as well as mass adoption. But in them we see the building blocks of a new and different way of structuring our online lives that combines security and autonomy, a re-decentralization with respect for privacy baked in.

Electronic medical data could be stored in a similarly decentralized way, rather than on centralized servers run by government authorities. We would need backup processes and secure storage, but such a system reduces incentives for hackers stealing data for pay (by eliminating the honeypot) and also has the potential to reduce the impact of cyberattacks like the WannaCry worm. Our data would no longer be concentrated in the hands of a few key players, with access points that can be used to surveil or force us to pay a ransom. We could share only what we wanted to share, on our own terms, and keep our information up-to-date under our own control.

These kinds of decentralized systems require planning. Unlike updating some kinds of personal information, it is critical that medical data, for example, be verifiable—to limit doctor-shopping for drugs, for example, or to get an accurate picture of events after a particular treatment or a medical mishap. Blockchain technology has great potential in this regard. Blockchain is a digital ledger that records transactions, without storing that record in a central place. Instead the record or ledger is dispersed and saved in multiple places. Blockchain involves creating a chain of data transactions (or blocks), and the evidence of the chain is distributed across numerous computers. The upshot is that there is no central gatekeeper or repository of knowledge, while the distributed nature of the ledger makes it theoretically impossible, and at least immensely difficult, to tamper with.

Perhaps most famously, blockchain is the technical basis of Bitcoin. Because blockchain removes the need for a central ledger (in this case, a bank) by devising a distributed ledger of transactions, it is possible to create a currency without a state. But cryptocurrency is not the only potential application of this technology. Certainly there has been some over-hyping of both bitcoin and blockchain in recent years. But there still remains plenty of scope to use the latter for other kinds of transactions and record-keeping systems.

Provided we can find ways to protect patient anonymity and confidentiality and obtain informed consent, it is possible to imagine medical records being stored in a decentralized way that is also verifiable. Various combinations of these technologies could also allow people to perform many other activities online that involve giving access to personal information. The information could be parceled for sharing with banks and insurers, for example. In other words, digital technology has the potential to give people power over their own data in ways that are accountable and effective.

The WannaCry worm created havoc in numerous universities, health systems and workplaces globally. Nearly a quarter of a million computers were affected in over 150 countries. The experience spotlighted the risks of centralized data on an enormous scale; it is urgent to plan and build systems that are resilient against these risks. Information systems that protect privacy—that give people control over their data—improve security for everyone. We have the technology to begin designing networks that are far more impervious to attack.

So why haven't these gone mainstream? Part of the reason at least is that the proliferation of technologies like decentralized data pods and blockchain undermine the ability of technology capitalism to monitor and profit from our digital behavior, and they limit the capacity of governments to tap into this infrastructure for their own surveillance purposes. In other words, this kind of data diffusion challenges the two central repositories of power in our society: capital and the state. The re-decentralization of the web may be a technological design issue, but it will only be achieved if we understand it as a political objective.

Here we encounter a "classic Fanonian theme" as described by Gordon: "There is no reciprocal respect without confrontation." In the words of Fanon himself: "You do not disorganize a society ... if you are not determined from the very start to smash every obstacle encountered." It is critically important to disorganize the society currently built around digital doppelgangers and segregated marketing. This is unlikely to happen of its own accord. "We do not expect this colonialism to commit suicide," wrote Fanon. "It is altogether logical for it to defend itself fanatically." In such circumstances, relying on the benevolence of state and capital to restructure digital society is a mistake. As Fanon concluded: "It is the colonial peoples who must liberate themselves from colonialist domination."

In Fanonian terms, the end of racial prejudice opens people from diverse backgrounds to the possibility of "actually becoming brothers." By making their existence known and visible, colonized people create conditions for reversing their erasure, which is the exact kind of problem we encounter in biased algorithms. Data discrimination, abstract identification, and categorization according to secretive rules all create situations where people undergo arbitrary prejudice and unnecessary suffering. Even after we open the black box algorithms and we know what the rules are, we need to generate the right inputs, find ways for the oppressed to exist in digital society and have control over their data so it accurately reflects their sense of self. By making this kind of demand, "the two cultures are able to confront each other and enrich each other," Fanon wrote. "Universality lies in the decision to support the reciprocal relativism of different cultures once irreversibly excluded by colonial status." Fanon saw confrontation as begetting a form of reciprocal respect, a universality that respects diversity.

What does it look like, to generate data that confronts prejudice and relieves suffering? An NGO called Transparent Chennai is a good example. It uses digital technology to map slums in India, including the public services available to the inhabitants. Prior to this, these residents were invisible to elected representatives and policymakers. Through a combination of open source technology and practical follow-up on the ground, activists "revealed" the existence of half a million people. It created a compelling case for improving infrastructure for a population that had been systematically ignored by public authorities. Transparent Chennai is one of many projects engaged in grassroots digital mapping, with a focus on empowerment rather than commercial gain. OpenStreetMap relies entirely on volunteers to create and provide free geographic data, such as street maps, to anyone. The work of the platform has assisted in mapping the sites of humanitarian crises, meeting a need in a way that could never be achieved by the private sector. Digital technology makes it easier than ever to generate large bodies of high-quality data; and the more participatory the process becomes, the better it will be.

Another field with huge potential is the sharing of health data. Computerized medical data allows researchers to collate and link data points in medical records all over the world. It means that obscure conditions can be identified more easily and unexpected successes in treatment more widely shared. It allows for better management of complex health conditions. But the risk is that such data can be used against those who have contributed it. Discrimination on the basis of genetic data in the provision of services is already illegal in many places, but regulation needs to keep pace with technological change. For example, in the United States, employees can often reduce their health insurance premiums if they participate in genetic testing programs, and this involves handing over their health information. Under proposed legislative reforms, employers could ask employees to disclose research information—which they may have obtained privately or through participation in studies—and penalize them for refusing to do so. This would create a disincentive to participating in clinical studies and to pursuing this kind of information privately (which may open up options for preventative care). It illustrates how the sharing of medical data will always produce better outcomes under socialized forms of medicine, where sharing sensitive data does not jeopardize the continuation of coverage or affect its cost. But regardless of context, it is critical to establish robust protections against the many kinds of discrimination that may arise from disclosing such data, particularly in the provision of services like insurance.

Digital infrastructure, when deployed to share data collectively rather than for private gain, can allow social diversity to become a strength rather than something to be erased or flattened. Crowdsourcing mapping data shows how giving people control over their own data and allowing it to inform open source projects can be empowering. Sharing of medical information also means that collectively we will benefit from improvements in treatments in exponential ways, but only if there are safeguards against this information being used to people's detriment. Devising methods for sharing personal data safely will mean that algorithms that are used in a wide range of settings will improve over time. We must embrace the possibilities of digital technology by making political demands around accountability for algorithms and nondiscriminatory treatment of data. Fanon observed this in how Algerians used radio: "Algerian society made an autonomous decision to embrace the new technique," wrote Fanon, "and thus tune itself in on the new signaling systems brought into being by the Revolution." We need to build and tune in to a new signaling system made by people committed to revolutionizing the web.

Early in his first major work, *Black Skin, White Masks*, Fanon hangs his inquiry around the question: "What do blacks want?" By so doing, he is tugging at our thinking around the experience of living a life, that is not as we choose. Seeing yourself through the eyes of those who oppress you—who treat you as an object or do not see you as a person at all, in the case of a nonwhite Algerian in French-controlled Algeria—can make answering this question difficult. How do you separate your own desires from the hatred heaped upon you, which you invariably heap upon yourself? It requires us to think about who controls us, who writes the history of our sense of self and tries to determine its future. If such control was stripped away, what is it that we would want? For the person who has been colonized or oppressed, there is a split between a subjective desire (what we might want) and reality (what our social structure tells us we want).

Our engagement with digital social spaces can create similar conundrums. We often use these platforms to seek connection and escape, only to find ourselves chained to the logic of gratification, buying things we cannot afford, spending time we do not have, becoming people we do not feel we are. Social media platforms pull our preferences to the fringes, by playing on our desire for drama and confirmation bias. This mutates our sense of self but also our capacity to relate to one another.

But it is not a case of our true, base natures manifesting themselves. The reality is more complex. There is something addictive about this experience. Richard Seymour suggests that we do work for these platforms, by publishing our experiences, but not in the form of waged labor. "Rather than being paid to write, as you would be if your content was published in traditional media formats, you are offered gratifications of the self." The outcome is that

> social media engages the self as a permanent and ongoing response to stimuli. One is never really able to withhold or delay a response; everything has to happen in this timeline right now, before it is forgotten. To inhabit social media is to be in a state of permanent distractedness, permanent junky fixation on keeping in touch with it, knowing where it is, and how to get it.

This "legalized crack" (as former Facebooker Antonio García Martínez describes it) makes it increasingly difficult to escape the compulsion to live life online, where nothing is designed to help us disengage. Our current modes of interaction encourage spending time on-device because it is a metric of value to corporations trying to sell us things. This impacts how we relate to each other in social spaces online. Rather than blame some kind of amorphous notion of human nature or some uniquely terrifying quality of screen devices, we would be wiser to apply Natasha Schüll's conclusions about electronic gambling (see chapter 1) and see this as arising from a dynamic interaction between device and person. Devices can be designed differently, to allow us to use them in ways that are fulfiling and functional.

Digital self-determination, then, should include the freedom to engage with digital life consciously and deliberately. We exist as individuals who are part of a collective, and creating the conditions to enjoy an autonomous digital life has to find ways to chime with this communal element of the self. We need to design online experiences that give space for human interaction, give breaks from mindless consumerism, and offer self-possession rather than escapism.

In his undeniably optimistic conclusion to *Black Skin, White Masks*, Fanon dares hope "that the tool never possess the man." Both blacks and whites

> must turn their backs on the inhuman voices ... in order that authentic communication be possible. Before it can adopt a positive voice, freedom requires an effort at dis-alienation ... It is through the effort to recapture the self, and to scrutinize the self, it is through the lasting tension of their freedom that men will be able to create the ideal conditions of existence for a human world.

Fanon's writings are a call to action, but they also frame an objective for action. Our efforts to recapture ourselves need to begin with us, but they also must be part of something bigger. Fanon was always oriented toward externally directed activity, rather than individual reflection alone. We need to find ways to create this opportunity in digital life.

The capacity to consciously and deliberately engage with digital life sits on the border between the self and the collective, and it is predicated on a sense of belonging. Digital self-determination requires care to ensure that digital spaces are welcoming and inclusive. The open nature of the web is such that how it regulates, or abstains from regulating, speech has been of central concern in many social spaces. An obvious distinction can be made between Facebook, which opts for a relatively heavy-handed, if somewhat inconsistent line on content management, and Twitter, which has for most of its history refused to censor, making it a "honeypot for assholes," as one former employee put it. Facebook's guidelines for content moderation appear to be highly complex and at times contradictory. Meanwhile, Twitter's public share offering struggled for a period, in part owing to the fact that, in the later words of a former CEO, "we suck at dealing with abuse." This problem has also reportedly thwarted attempts at the sale of shares in the platform.

Critics of digital society often claim that digital technology is a black mirror, reflecting the truth of human nature. Our online spaces are ghoulish and sadistic because this is our natural disposition. But Fanon reminds us that how we conduct ourselves is often less about our inner desires and more a reflection of how society is organized. In Algeria, Fanon witnessed this phenomenon firsthand in the form of various pathologies. "Imperialism," he wrote, "which today is waging war against a genuine struggle for human liberation, sows seeds of decay here and there that must be mercilessly rooted out from our land and from our minds." In *The Wretched of the Earth*, Fanon set out some examples of decay, including rape, random homicidal impulses and torture. "Honor, dignity and integrity are only truly evident in the context of national and international unity," Fanon observed. "As soon as you and your fellow men are cut down like dogs there is no other solution but to use every means available to re-establish your weight as a human being." The experience of occupying digital spaces is hardly the same as fighting a war of national liberation. But it is easy to identify with Fanon's description nonetheless, to appreciate how life online can be a dehumanizing experience and to know the power of the desire to re-establish one's weight as a human being.

Having an online space that is welcoming and not exclusionary is a concern for many people. We need to create systems that help manage the problems of trolling and bullying, particularly in respect of women. Too many of these platforms fail to manage abuse; indeed, abuse occurs in ways that fully authorized or expected by the platform. Researcher Cade Diehm argues for a rethink in how we approach user design, to think beyond the experience of the ideal or even typical user and instead start to anticipate the ways in which platforms and products can harm vulnerable users. This calls for a design approach that intersects with security research. "The goals of the two fields are diametrically opposed," he writes. "Design is to create the best possible experience for a user, security is to create the worst possible experience for an attacker." In this way, a security lens can help improve design by anticipating unwanted uses of the platform and detering them before they happen, with the aim of minimizing abuse. In doing so, technological design can enhance our sense of belonging and give us the power to determine how we experience digital life. Plenty of designers have offered concrete suggestions to address some of these issues. The problem is not the shortage of ideas but rather a lack of appetite for them.

Perhaps it is time for public ownership of a platform like Twitter, to allow users to have a say in how it is run. The idea of nationalizing railways and other utilities is now almost mainstream in some social democracies. Companies like Twitter, which built itself on the back of public investment in technology and now provides a form of public infrastructure, could be the object of a similar demand. Such a campaign could bolster the call for improved user experiences that facilitate exchange and debate but safeguard against abuse and prioritize user control.

Building open and inclusive platforms is urgent and necessary; it is required to facilitate full participation of people online from diverse backgrounds. It is not a total answer to the problems of discrimination and division, but it is a critical component of any progress toward ending them. As is the case in so many contexts, increasing the input of people from marginalized communities into discussions about these topics will contribute to a better understanding of the issues and the generation of creative ideas to overcome them.

In closing his final work, *The Wretched of the Earth*, Fanon talked about Europe as a place where the rights of man are respected but only through the violent abrogation of the rights of others not allowed the full designation of European political personhood. Colonialism was the ugly underbelly of the European project of enlightened reason, inextricably linked to crimes including division, class stratification, racial hatred, exploitation, slavery and "the bloodless genocide whereby one and a half billion men have been written off." Fanon warned his African comrades against the temptation of trying to build another Europe in Africa.

Liberal regimes of rights have rightly attracted criticism for being hollow. They allow rich and poor alike to sleep under bridges, rich and poor alike to give away their data and be manipulated by black ops marketing or surveilled by bigoted police, to iterate on Anatole France's indictment of law. Fanon's understanding of Europe was focused on its history as a colonizer and the experience of those who were the objects of colonization. But we can heed his warning today in a quite different context. Rights are nothing without duties. We should not expect the powerful to respect individual rights but collectively create systems designed to impose that respect as a duty. For rights to be meaningful, we need a system that holds power to account.

As we will see in the following pages, the idea of Europe and its supremacy overshadowed not just African lands but also many others around the world. It robbed people not just of self-determination but also of the possibility of other ways of existing in the world. "The belief in European supremacy legitimised the violent theft of all things Aboriginal—our lands, our lives, our laws and our culture," reflects the Australian Aboriginal scholar Irene Watson. "It was a way of knowing the world, a way which continues to underpin the continuing displacement of Aboriginal peoples." Dismantling this idea will require thinking about how we can implement alternative ways of governing not just ourselves but also our commonly held resources, a topic explored in the next chapter.

In Algeria, Fanon witnessed firsthand how oppressed people took back technology from the control of colonizers and repurposed it to speak the first words of a new nation. "The nation's *speech,* the nation's spoken *words* shape the world while at the same time renewing it." The technology of the twenty-first century gives us a similar opportunity today. Critically for Fanon, the way that this happened was through struggle. "The same time that the colonized man braces himself to reject oppression, a radical transformation takes place within him which makes any attempt to maintain the colonial system impossible and shocking." This is the potential of digital self-determination today.

10

The Digital World Is an Environment That Needs to Be Cared For

Ancient Forms of Governance Hold Relevance for Modern Infrastructure

The Māori *iwi* of Whanganui, in the North Island of New Zealand, fought for the recognition of their ancestor for nearly 150 years. But that ancestor is not a person or a name or a family. It is the Whanganui River, the third-largest in the country. As part of a settlement under law in 2017, the river was recognized as a legal person, meaning it had the rights and liabilities that derive from that status. "From our perspective treating the river as a living entity is the correct way to approach it, as in indivisible whole," said the lead negotiator for the iwi (or tribe), Gerrard Albert, "instead of the traditional model for the last 100 years of treating it from a perspective of ownership and management."

The Whanganui case was not the first settlement of this kind. In 2012, the Tūhoe negotiated a similar arrangement over their lands in Te Urewera National Park. As well as financial redress from the New Zealand government, the Tūhoe steadfastly sought autonomy over the lands. They were not looking for rights or concessions but self-governance. There are also precedents for this kind of legal status in other parts of the world. The Colombian Supreme Court of Justice declared in 2018 that parts of the Amazon possessed legal rights to protection.

The Māori approach to land is deeply distinctive from Western conceptions of ownership. They do not view the natural environment as a resource to be exploited, but understand human beings as just one part of the universe, living together with land, rivers and sea. For the Tūhoe, this affinity takes the literal form of kinship—one of their ancestors was the offspring of a union between mountain and mist. Everything about the Tūhoe is "entangled in this place we call Te Urewera," explained the chief negotiator, Tamati Kruger.

Treating a river or tract of land as a person may seem curious at first, but Western legal systems too assign personhood to a variety of things. If companies and trusts can be legal persons, it is not difficult to imagine granting similar status to a river, especially given the importance of a healthy environment for our own physical well-being. But it remains novel legal territory in Western law. The Tūhoe settlement was the first of its kind, and the Whanganui settlement represented an end to the longest-running litigation in New Zealand's history. But even though these cases took a many years, the Māori position remained steadfast. As the representative for the local area of the Whanganui, Adrian Rurawhe, put it: "It's not that we've changed our worldview, but people are catching up to seeing things the way that we see them."

Indigenous communities, including the Māori and equivalent populations in Australia and Canada, have a lot to teach us about governance and the natural world. While there are certainly differences across these peoples, there are also commonalities. Colonial projects across the globe were consistently intent on eradicating these approaches and replacing them with their own. To put it in more positive terms, these societies had sophisticated systems of law prior to colonization, though they are rarely afforded the respect they deserve in mainstream histories. In his fascinating examination of the precolonial Aboriginal Australian economy and society, Bruce Pascoe makes an urgent call for an examination of what he claims to be "the longest lasting pan-continental stability the world has ever known," to learn how "that cooperation was wrought without resort to the physical coercion and war common in other civilisations." Melissa Lucashenko, an Australian Aboriginal woman, argues persuasively that "there is overwhelming evidence from Aboriginal oral history, as well as from anthropology and written history, that Aboriginal adults traditionally saw the world in highly egalitarian terms compared to their European contemporaries." For thousands of years prior to colonization, she argues, they negotiated a "structured peace" across vast territories and multiple language groups. She concludes: "If that isn't a democratic outlook, it's hard to know just what is."

There are real questions about the extent to which Aboriginal laws and customary practices can be meaningfully or effectively synthesized with dominant Western legal systems. The thinking about Aboriginal governance and decolonization is understandably diverse, and there are complexities around the language of sovereignty and self-determination that often reflect disagreements on how power is exercised through law and politics. My object here is not to resolve these questions or to suggest that the thinkers cited here are congruent or even aligned, but rather to uphold these ideas as points of discussion that might helpfully inform our approach to the governance of collective resources.

In the preceding chapter, we saw how Fanon reckoned with personal identity under colonialism. According to Tamati Kruger, the Tūhoe have their own definition of this idea, which he expresses poetically: "*Mana motuhake* [self-government/determination] is the ability to live your own dreams, rather than being forced to live the dreams of others." This profound idea evokes more than an individual pursuit, focused on a sense of self: decolonization has a broader history and practice that regularly plays out in Indigenous communities around the world fighting for their land. Fanon's warning about Europe, and the need to avoid reproducing that society in Africa, is constantly finding new forms of expression in places that are coming to grips with the legacy of colonialism. Stories like that of the Whanganui River and the Tūhoe show how former colonies can grapple with their past in ways that are inclusive and constructive. Indigenous ideas of governance are not historical relics or anomalies; they hold valuable knowledge for how we can move forward in modern society, with relevance to various problems arising in the course of the digital revolution.

A Mohawk writer and professor from Canada, Taiaiake Alfred, has analyzed how the "conquest mentality" created a structure of relationships between people and the land "that is ultimately destructive to everyone and everything involved." Understanding and appreciating Indigenous approaches to governance is an important task, not just as part of an attempt to right historical wrongs or as a way of coming to terms with the cultural significance of land. Rather, Alfred believes, it is necessary to save the world from the logic and drive of capitalist development, which is lacking an ethical framework or principles of sustainability. Unlike capitalism, Indigenous concepts of governance explicitly place limits on the ideas of growth and exploitation. The Australian Aboriginal scholar Irene Watson explains things another way: "We live as a part of the natural world; we are in the natural world. The natural world is us," she writes. "We take no more from the environment than is necessary to sustain life ... Settler societies have lived on Nunga [First Peoples'] lands and taken more than is needed to sustain life and the result, as we know, is the depletion of *ruwe* [the territories of First Nations People] and the exhaustion of natural resources." Unless something changes, we can expect a climate apocalypse with devastating consequences.

These principles have relevance to how we discuss the regulation of collective digital spaces. This is not a bid to appropriate these ideas but rather to argue that we have much to learn from how these societies maintained peace with each other in the context of managing shared resources. It is an attempt to avoid assuming the centrality of European traditions of law and governance, and to admit how these assumptions have crowded out other perspectives. As Watson puts it, "very few ever consider asking the question: what laws existed before colonization or, what happened to those systems of law?" When we ask this question, we start to think about how First Nations' laws hold "the potential for the future growing up of humanity." Relatedly, we can also draw from more recent movements for environmental protection to build on their strengths and anticipate their weaknesses. The Internet is a resource—both a history and a future—that we ought to hold in common. It crosses national borders and provides collective value and wisdom. It demonstrates the power and value of interconnectedness and interdependence. But it is also vulnerable to exploitation and the insidious consequences of privatization. It is an environment that needs to be cared for.

It is also the environment in which we increasingly live. The environment is relevant to the broader purpose of this book in two ways. First, it serves as an analogy. Like the natural environment, the digital network is a shared infrastructure that we all utilize, and how it is governed will have implications for how we relate to each other socially. If we think of the Internet as something physical that must be regulated and protected in the material world, there is a lot to learn from the history of environmentalism and the way in which Indigenous communities relate to land in places like New Zealand, Australia and Canada. By regulating the network in ways these communities value, with fairness and respect, we can avoid this common resource being used to accumulate wealth for a minority—which, in the environmental context, is giving rise to devastating climate change.

There are limits to the environmental analogy: most obviously, the digital network does not hold spiritual significance in the way that the land and environment does for Indigenous people. Digital infrastructure is a human creation. The analogy is not aimed at belittling this in any way but, as with other chapters in this book, draws inspiration from alternative relationships to the physical world and its shared resources. As Alfred points out: "Scholars of international law are now beginning to see the vast potential for peace represented in indigenous political philosophies." Scholars of the digital society would be well advised to do the same.

Second, there is a more consequential relevance: our capacity to address climate change is linked to the extent to which we can make the most of the digital revolution. Digital technology has the potential to maximize our use of limited material resources in all sorts of ways, including in relation to food supply, energy production and resource management. Insofar as we are learning from the frightening prospect of catastrophic climate change, we might apply that learning to how we approach governance of the Internet. Peter Frase talks about this in *Four Futures* (2016), which he frames around the twin anxieties of ecological disaster and the automation of work. These two issues, he argues, are "fundamentally about inequality":

> They are about the distribution of scarcity and abundance, about who will pay the costs of ecological damage and who will enjoy the benefits of a highly productive, automated economy. There are ways to reckon with the human impact on the Earth's climate, and there are ways to ensure that automation brings material prosperity for all rather than impoverishment and desperation for most. But those possible futures will require a very different kind of economic system than the one that became globally dominant by the late twentieth century.

How we discuss and make decisions about the digital society has urgent relevance for our ecological future. We can no longer postpone the task of managing our resources differently from the status quo.

Tamati Kruger recalls a moment in the negotiations between the Tūhoe and the New Zealand government when talks broke down. He realized that the prime minister then, John Key, had misunderstood what the Tūhoe were seeking:

> Ownership was his obsession, not ours. So now we don't use that word. It's not a Tūhoe concept anyway. In the course of these negotiations, I've had to study this European understanding of that word "ownership" and

where it came from. Ownership is the proof that something is yours to sell. So it is more about how to rid yourself of something, to gain material benefit from it, than to preserve and keep it.

Alfred echoes this sentiment, noting that "sovereignty" is an inappropriate political objective, as it is based on an idea of dominance and rights, buying into the colonial mentality. His aim is to rebuild cultures of mutual respect, both for each other and our surroundings, drawing on the work of Indigenous activists dedicated to bringing these cultures to the fore. "Our goal should be to convince others of the wisdom of the indigenous perspective." In the twenty-first century, we need to create a digital environment that is not owned by anyone or any entity but is preserved and protected for a shared future, based on a culture of mutual respect. We need to start thinking about the Internet as a landscape that creates the conditions in which we live, as a shared responsibility that we contribute to and draw from.

Trevor Paglen photographs invisible things. Better said, his interest lies in the point at which the invisible becomes visible, where the imagined becomes material. Paglen has spoken about how we often use a range of mystifying metaphors to describe the Internet, when not trotting out hopelessly clichéd images of digits and light, like a bad version of the Matrix. From the perspective of intelligence agencies or service providers, however, the Internet looks rather different. It is an infrastructure, made up of cables, nodes and choke points. Major continents are strung together across seas and oceans. Fiber optic cables fuse cities across land. Maps of this infrastructure plot out how we communicate. Its components are all owned and controlled by specific groups of people. This sense of ownership—together with the right to gain material benefit from it, as Kruger might put it—is something we often forget when we picture the Internet.

Paglen's photos include a collection of underwater scenes, murky and unremarkable, save for a cable running through them. This is the Internet in concrete form. It is shocking to see something so essential to our lives today looking rather paltry in a dreary vista of the ocean floor. The cables are surprisingly slim, seemingly quite vulnerable. Yet much of our daily traffic of human knowledge, life-giving infrastructure and communication travels at the speed of light down the spines of these dismal sea snakes. Just as the health of the Whanganui River is connected to human vitality, these cables are critical to our systems of food and energy production, knowledge, personal safety and cultural heritage. How this system is governed has an impact on our individual and collective health and well-being.

Paglen's photos encourage us to think of the Internet as connected objects rather than a service or a space or some more fanciful metaphor (perhaps most notoriously, yet quite accurately, "tubes"). They help us appreciate the hard truths of geography as applied to the Internet. They force us to consider how our digital society exists in material terms, and how it is governed. These cables did not come from nowhere; they were laid and are controlled by people. As professor Philip N. Howard puts it, this physical reality reflects a political world order—what he calls *pax technica*—a pact between big technology firms and government:

> Undersea trunk cables connect continents. Private businesses own and maintain those cables. Those businesses have lobbyists who try to acquire bigger and bigger contracts from governments. A simple way to describe this relationship is to say that politicians made investments in surveillance firms, and surveillance firms made investments in politicians.

The chief material beneficiaries of the Internet are businesses, keen to promote commerce using digital technology, and those who run the surveillance state. This *pax technica* is a political idea that takes physical form in these cables, but all too often it remains obscured from view and therefore easy to forget.

If we dive a little deeper, the history of the Internet backbone tells us how its physical reality has been shaped by the interests of private industry and government, but also that it need not have been this way. The pipes and switches that connect us today were originally built with public funds, for public use. It was a political decision to hand them over to commercial entities, which in turn comply with the demands of the surveillance state.

How did this happen? In the early days of networked computing just after World War Two, various people and organizations started to see the value of connecting large computers. A number of different networks developed over this time, both publicly funded (through the US military) and commercial equivalents. The first and most important network was not built by business or government but—appropriately—by a collective. Michigan State University, Wayne State University and the University of Michigan together conceived the Merit network as a method for communicating between themselves for research purposes. It turned out they were really good at it.

The Merit network received some initial funding in the late 1960s from the Michigan state legislature and the National Science Foundation (NSF), an independent federal agency to promote the progress of science, improve national health, prosperity, and welfare, and secure the nation's defense. Merit was one of several similar projects the NSF had under way, including ones dedicated to expanding networks and building national infrastructure that could eventually roll out country-wide access to the Internet for everyone. Throughout the early 1980s, the NSF built connections between five funded supercomputer centers and a high-speed backbone that connected over 200 regional and campus networks. Together this was known as the National Science Foundation Network (NSFNET) or the Internet backbone.

Eventually, as traffic and users increased, maintaining the Internet backbone became too much for the NSF, and it solicited assistance for managing this national network in 1987. Merit was up to the task and won the bid. Indeed, against the prevailing expectations of universities, as compared to commercial or military operations, it performed exceedingly well. Merit ran the NSFNET for seven and a half years, during which time, according to a former president of the organization, Eric M. Aupperle,

> backbone network traffic increased nearly a thousandfold, reaching almost 100 billion packets per month. The number of Internet networks announced on the backbone grew from a handful in 1988 to 50,766 in April 1995, of which 22,296 were non-US networks. The number of countries comprising the Internet grew from 3 to 93. The extraordinary success of the NSFNET project was the dominant factor in converting the ARPANET and early NSFNET research and academically focused Internet into today's worldwide commodity Internet phenomenon.

Aupperle concluded, just over a decade after the project ended: "Merit and our partners are proud and pleased to have been a key part of this exciting transition."

So why did this project come to an end if it was working so well? The answer is politics.

The NSFNET emerged at a time when large social investments were increasingly considered politically passé. The neoliberal agenda was gaining force on both sides of the mainstream partisan divide, with a renewed and vigorous faith that the market could deliver services better than the lumbering bureaucracies of government or public entities. The government's job, according to the growing consensus among the political class, was to regulate markets only minimally and allow the private sector to flourish, privatizing services that were previously considered the purview of public institutions. This was reflected in the approach taken to the Internet backbone. Commercial traffic was not permitted on the NSFNET. The idea was that this prohibition would create incentives for commercial networks to develop concurrently. This reflected a long-term plan for the network to be commercialized, and presumably expand, so that researchers would ultimately piggyback off an expanded commercial Internet.

Everything went exactly to plan. Commercial providers sprang up, and over the course of the next decade the US government supervised the complete privatization of the Internet backbone network. There were no meaningful regulations imposed on the use and development of the Internet infrastructure. The NSFNET was eventually defunded.

This represented the privatization of a public infrastructure project, with consequences for our understanding of the collective. The government imposed a sense of ownership on something that was actually the subject of a common responsibility, and by the same token it disavowed a more social and relational approach to this piece of shared infrastructure. As Watson points out, "relational philosophy is embedded in Indigenous knowledge systems; knowledge belongs to a people and the people belong to a landscape." This approach elevates a sense of responsibility that is lacking in the neoliberal agenda that seeks the privatization of public resources. "Indigenous knowledges, unlike those of Europe, carry obligations and responsibilities, such as custodial obligations to ruwe that bind future generations," writes Watson. When lawmakers sold off our digital infrastructure, they were motivated by a philosophy that presumes that market-based systems of property and profit are superior for our collective development. They short-changed us all.

The outcome of this has been an Internet that operates at a substandard level in the continental United States. Despite being the place where the Internet all began, the United States cannot match other developed nations on speed, coming in tenth on average. Most acutely, speeds in rural areas remain slow, with few firms in the private market competing over connections. According to the Federal Communications Commission, "many Americans still lack access to advanced, high-quality voice, data, graphics and video offerings, especially in rural areas and on Tribal lands." The wholesale privatization process effectively created an oligarchy: a select few companies were able to consolidate commercial advantages at the time of privatization and create barriers to meaningful competition. The fragmentation of the backbone created a situation where large providers cut deals to manage traffic across networks. These are unregulated, secret agreements, made without any accountability.

As a result, newcomers seeking to enter the market do not compete on a level playing field, and the position of established corporations has been effectively consolidated. Companies control how traffic flows through the tubes, with no oversight, standards or performance requirements imposed by the government.

It is deeply concerning that we do not even have a map of the Internet backbone produced by a public authority. Thankfully some academics shouldered this task, but it took a dozen of them four years to produce, painstakingly compiled from multiple sources. They concluded that this kind of mapping has strong potential to improve efficiency and robustness, protect against natural disasters, and guard against criminal attacks. But more generally, it is also essential to Internet governance. It is revealing that this resource had to be produced by academics rather than the industry or government.

This is an example of the accumulation of technological infrastructure by industry, discussed by Luis Suarez-Villa in his examination of the nature of capitalism and digital technology. "The rapid accumulation of public technological infrastructure during the second half of the twentieth century," writes Suarez-Villa, "subsidized the power of corporatism, allowing it to become more effective in extracting value out of new knowledge." Value in this context does not mean for society or humanity in general, only for business corporations.

In short: public funds paid for the invention and development of a key piece of infrastructure. Then, in line with neoliberal economic philosophy, the government decided to outsource this to the market for the purposes of scaling up, for no good reason. The government could have easily continued to fund the backbone and its entire regional expansion—an ambitious program for sure, but no more so than a national rail or road system, and, moreover, it was something it was already doing. There is every chance this would have resulted in a faster and more efficient network. Some of the best aspects of the Internet as a network, its capacity to connect people in conditions of openness and neutrality, without incorporating tools to control its users, stem from its nonproprietary origins in academic institutions rather than corporations. At the very least, the privatization of NSFNET represented an opportunity to incorporate social concerns into the functioning of this critical piece of infrastructure. But all these opportunities were lost. No meaningful requirements were imposed on the industry, not even the right to produce a map of its existence—the government simply gave it all away, ideologically convinced that the market delivers services better than governments or public institutions. Our current reality flatly contradicts this.

That old chestnut about the value of the market needs to be jettisoned once and for all. As Ben Tarnoff notes, there was nothing inevitable about the privatization of the Internet backbone. "It reflected an ideological choice, not a technical necessity," he writes. "Instead of confronting critical issues of popular oversight and access, privatization precluded the possibility of putting the Internet on a more democratic path." Professor Howard has pointed out how technologists and venture capitalists "love comparing the current internet of things to the Wild West—lots of people doing crazy, creative things on the frontiers of computer science and engineering." The analogy is dismayingly apt, in fact: just as the "frontier" in American history was based on the dispossession of Indigenous people for private gain, so too the great privatization of the Internet backbone represented depriving people from access to knowledge, community and empowerment in the interests of a select few.

As part of her analysis of Aboriginal law, Watson recounts the story of the greedy frog:

> The frog story tells of a time when a giant frog drank up the water until the land was all dried up and in drought. To survive, the collective of animals agreed it was necessary for the frog to release the water it had drunk back on to the land. They decided that the strategy most likely to succeed was to make the frog laugh; by laughing the frog would release the water. After many attempts at humouring the giant frog, the animals succeeded and the frog let forth a large laugh and with it released the water back on to the dry lands, filling up lakes, creeks and riverbeds. As a future precaution, the animals then decided that they would prevent the

> event occurring again by reducing the power of the frog. So, instead of there being one giant frog, many smaller frogs were created and the frog was never again in a position of power to monopolise the land's waters.

The story highlights the dangers of the centralization of power, especially when it governs the distribution of life-giving resources. We could apply the lessons of this story to our thinking about the governance of digital infrastructure.

Some alternative ways to create an ecosystem of digital infrastructure focus on public ownership at a local level. In spite of the wholesale privatization of the Internet backbone, examples of public projects afford a glimpse of how things could have been done differently in the past, and how they might be done in the future.

Local government authorities across the United States have begun investing in digital infrastructure themselves, with surprisingly positive results. MuniNetworks.org, a project of the Institute for Local Self-Reliance, has collated data on the more than 110 communities in twenty-four states with a publicly owned network offering at least one-gigabit-per-second services—twenty-six times faster than the national median. Chattanooga, Tennessee, is one example. An agency owned by the city, the Electric Power Board, has set up its own network with a government grant. The network offers higher speeds than any of its private-sector competitors, making Chattanooga the first city in the United States to have a citywide gigabit-per-second fiber Internet network at a reasonable price. Media reports note that "the city has established itself as a center for innovation—and an encouraging example for those frustrated with slow speeds and high costs from private broadband providers." Tarnoff reckons that these kindts of projects could be the first steps toward a "popular movement to reverse privatization."

Such initiatives are our future. Time is up on the era in which the Internet is owned by an oligarchy; the experiment with privatization, with "the impulses and the imperatives of capitalist development" as Alfred called it, must come to an end. We must demand the right to see the pipes and switches of the Internet, to map its physical existence so that we can plan how to expand and strengthen it, through good design and ambitious public investment. We must demand the right to take shared public responsibility for the Internet backbone and its tributaries, so that we can, in turn, take shared responsibility for own participation in digital society. The Internet backbone was a creation of public spending; the public has a claim over it. Governments should socialize the backbone, bring it under the collective authority of a democratically accountable authority, and allow autonomous local investment programs to deliver fast and efficient connectivity across the globe. To allow giant, greedy frogs to exist risks robbing us of a future in which we can share in abundant, essential resources collectively and responsibly.

It can be easy to dismiss these traditions of knowledge and rule-making when they sit outside accepted paradigms. But we risk missing important insights and foreclosing alternative possibilities when we take for granted that our current approach is universal. How can we find common ground with people from diverse traditions—with alternative ways of speaking, knowing and determining identity—without overriding difference? This is a critical question in the digital age, when forms of representation are increasingly applied to everyone.

Taiaiake Alfred argues that "the primary goals of an indigenous economy are the sustainability of the earth and ensuring the health and well-being of the people." The Indigenous relationship with land, contrary to the settler project, is one of partnership. This means that differences of perspective are recognized rather than ignored or overruled. Indigenous philosophical alternatives are based on values such as a "wish to preserve a regime that honors the autonomy of individual conscience, non-coercive forms of authority, and a deep respect and interconnection between human beings and the other elements of creation." As a result, Indigenous conceptions of politics resist homogenization. They "explicitly allow for difference while mandating the construction of sound relationships among autonomously powered elements." Western ideas of statehood are almost the opposite: rather than generating respect organically, from the bottom up, they "create a political or legal hegemony to guarantee respect." But this cannot last, says Alfred. "It is no longer possible to maintain the legitimacy of the premise that there is only one right way to see and do things."

What might this kind of equality, or respect for difference, look like in a digital society? When we discussed the idea of universal access to the digital network in chapters 7 and 8, it was framed around equality in the public spaces and platforms that facilitate democratic decision-making. It was a question of enfranchisement within systems of power, including political bodies and workplaces. But this universality is no excuse to flatten difference. Once we have equality of access to democracy in economic and public life, we need to think about respecting difference, allowing this to constitute a foundation for sustainable prosperity rather than the seeds of division. This will require that we pay specific attention to the building blocks of digital society, to turn our minds to some of the layers of the Internet above the pipe and switches, to take steps to ensure they enshrine respect for difference rather than entrench discrimination.

For example, neural networks will need to be trained in ways that explicitly acknowledge and address the problem of bias and allow space for divergence, conflicts and nuance. Word-embedding models illustrate the importance of this. These models operate by feeding huge amounts of text into a computer for the purposes of learning how words relate to each other in space. This is based on the premise that words that appear near one another within texts share a semantic meaning. These spatial relationships (or vectors) are used in natural language processing, teaching computers how to engage with us in human language. Word-embedding models are used all the time in many basic computer processes, the most obvious being the keyword search.

One way to demonstrate their functionality is by asking the model how words are associated with each other by analogy. For example, by reading a lot of text, a computer can learn that "Paris is to France" as "Tokyo is to Japan." The computer develops its own kind of dictionary by association, and words can be associated with one another in varying degrees of strength.

But this can create problems when the world is not as it ought to be—when the sense of universality undermines certain particularities. For example, researchers have experimented with one of these word-embedding models, word2vec, a popular and freely available model trained on 3 million words from Google News texts. They found that it produces highly gendered analogies. For example, analogies produced by the model for "man is to woman" include "computer programmer is to homemaker." Or "father is to mother" as "doctor is to nurse." This is not surprising on one level, as men are overrepresented in the ranks of computer programmers, and women are more likely to be nurses than doctors. But it also objectively reflects a bias that will be reproduced when we engage with the computer using natural language that relies on word2vec. It is not hard to see how this model could also be racially biased or internalize forms of bigotry toward other groups. Programming around natural language involves a huge amount of complexity. For example, neutral words are often associated with men (such as actor) and words associated with women often require an adjective to neutralize them (such as male nurse). Then there are metaphors and sayings (grandfathering a provision, keeping mum) that use gendered terms. It turns out that mapping language is messy and perpetuates all sorts of biases that are not always obvious.

The research found that these biases not only exist in the corpus that trains the model (the Google News text), they can be amplified during the process of language learning when using the word-embedding model. As the *MIT Technology Review* points out: "if the phrase 'computer programmer' is more closely associated with men than women, then a search for the term 'computer programmer CVs' might rank men more highly than women." When this kind of language learning is used in medicine, education, employment, policy-making and criminal justice, for example, it is not hard to see how much damage these kinds of biases can cause. Researchers on other projects have also encountered this problem in photo data.

Two image collections used in research projects were shown to have some troubling gender biases: "Images of shopping and washing are linked to women, for example, while coaching and shooting are tied to men."

This kind of problem is not unlike other biased algorithms we have discussed in this book, including Latanya Sweeney's personal encounter with her non-criminal record and the mistaken labeling of the Al Jazeera bureau chief as a terrorist by the NSA. Natural language or image processing is still in essence an algorithm, and the logic used to arrive at conclusions is created and refined by humans. In some ways it is more insidious, because the failures are difficult to detect—there is something deeply alien about how computers learn language, completely different to how we do it as humans, meaning the errors can be disconcertingly unexpected and easily missed.

But there is something else distinctive of natural language: all language is a metaphor; it involves a symbol that replaces the real. This is not a problem of "garbage in, garbage out," as it can be with other algorithms, because there is no alternative to language, no superior base of evidence or data to feed into neural networks. The language is reflecting the world as it is, and it is not the place many of us hoped it would be. Do we use language to seek to understand the world as it is (more women are in fact nurses) or how it ought to be (nursing should be a vocation free of gender stigma or association)?

There are other examples of engineered processes that flatten difference. Miriam Posner has spoken about this as part of her work in the digital humanities. She raises examples like mapping. The ever-popular Google Maps entrenches the Cartesian model of picturing space and sidelines other approaches, including those from Indigenous societies. Models for collecting and using data simplify concepts like race, disguising its nature as a socially constructed and ongoing process of self-definition. "Common types of data visualization," Posner contends, "are great for quickly conveying known quantities but terrible at conveying uncertainty or conflicting opinions." Adam Greenfield has cautioned against the tendency to assume that digital maps are "objective accounts of the environment." The reality, Greenfield argues, is that they are no such thing: "Our sense of the world is subtly conditioned by information that is presented to us for interested reasons, and yet does not disclose that interest." This is perhaps most obvious when observing how advertising helps to organize mapping technology on apps like Google Maps. But the program designers, for example, have also renamed whole suburbs in places such as Detroit and San Francisco, exercising this power without any meaningful accountability or transparency. It is unclear whether this is intentional or a function of the software design. The same problems arguably exist for almost any algorithmically determined form of representation.

These are layers of our digital infrastructure that have the potential to import inequalities and flatten difference unless we make the effort to identify and address them. We need to find ways to build into our computerized processes more space for nuance and conflict, more ways to define and respect difference, discarding the premise that there is only one right way to see and do things. We need to understand that the world is complex and diverse and draw on alternative approaches to how knowledge can be gathered and used—and this should inform how we represent it in digital settings. Computers and data can tell us how the world is, but we should use that learning to have a meaningful discussion about how it ought to be.

Professor Ben Schmidt offers some insights along these lines in his study of gender dynamics as they manifest in websites that collate reviews of professors. Analyzing this text using the word2vec mode, he draws an assortment of fascinating but curiously unsurprising conclusions regarding how people speak about male and female professors. Such as: "Students have a far more elaborate vocabulary to criticize women for being 'unprofessorial' than to criticize men." Or equally: "Students complain about the work their female professors assign: [not] because they have some other word for male workloads, but just because the workloads in woman-taught classes seem more worthy of comment." Schmidt points out that these kinds of programs do certain things, and we should be careful about ascribing to them qualities they do not possess. In a reversal of the classic quote from Marx, when it comes to word embedding, he says, "the point is not to change the world, in various ways, but to understand it." Word-embedding models allow us to appreciate the nuanced ways in which language influences our thinking, and enrich our understanding of the deep and complex ways in which gender (and presumably other forms of social categorization) influences how we engage with the world.

We need to embrace this possibility for understanding, so it can create meaningful ways of addressing bias and incorporating difference. Digital technology offers myriad ways to collect these understandings and put them to work. We can address the gender biases inherent in language and picked up by word-embedding models, and the researchers critiquing word2vec have suggestions in this respect. We can also build alternative ways of representing space, as well as lived experience on all sorts of online platforms.

In 1769, while visiting Tahiti, Captain James Cook met a man called Tupaia, from Ra'iatea in the Pacific Islands. Tupaia was a senior man and an expert navigator who ended up joining the crew of Cook's ship, the *Endeavour*. During his time aboard, Tupaia strove to communicate to the crew what he knew about the enormous space of Polynesia, a history passed down through generations. Communicating this knowledge was a weighty task: the settlement of Polynesia, spanning Hawai'i, Aotearoa (New Zealand) and Rapa Nui (Easter Island) was the last great premodern migration. In his history of GPS, *Pinpoint*, Greg Milner argued that "it is hard to overstate the immensity of this accomplishment." It remained something of a mystery as to how it was achieved, even to the expert navigator Captain Cook.

Tupaia crafted a map that encompassed Tonga, Samoa and New Zealand. In it he demonstrated how Polynesians, using astronomy and seasonal winds, achieved navigational feats that were beyond his captain's imaginings. Yet Cook appeared to miss the importance of this document in dispelling the mystery of how Polynesia was settled. Over time, Tupaia became alienated from the crew and died in 1770. The crew had found him to be "proud and austere," a social affront in racial terms. Cook himself seemed to find Tupaia disagreeable for failing to conform to the deferential posture expected of nonwhite people at the time. The historian Dan O'Sullivan suggests that Tupaia's self-confidence may have extended to geographical matters, "a cardinal sin" on an eighteenth-century ship. This disrespect for alternative perspectives meant that the complexities of Tupaia's knowledge were grossly underestimated and all but lost.

What other forms of knowledge have we lost as a result of privileging specific forms or representations? When everything is mapped by Google in a Cartesian framework, what nuances and alternative possibilities are excluded from our understanding of the world? Digital technology has the potential to accelerate our appreciation of alternative perspectives, but it requires an openness to and respect for diversity. Posner suggests we work toward "the thrill in capturing people's lived experience in radical ways, ways that are productive and generative and probably angry, too." She points out that this includes, for example, Aboriginal forms of picturing space using song lines rather than grids. The possibilities are endless, so long as we first agree that such biases exist and must be addressed by respecting positions of difference rather than glossing over them.

Just as it is for the climate, protection of the digital network is now essential for all sorts of everyday activities. Like the natural environment, it is a place of knowledge, enjoyment and self-exploration rather than just a resource for exploitation. The Internet was seed-funded through public money. It is the backbone of a huge amount of collaborative human effort and some of the most exciting developments in multiple fields of human endeavor. It should be returned to the domain of public ownership, through socialization and investment of public funds. It ought to be governed by rules agreed on transparently and in the interests of all users. Responsibility or accountability in the Indigenous sense, writes Alfred, involves a "requirement for universal inclusion and the maintenance of strong links between those charged with the responsibility of decision-making and those who will have to live with the consequences of their decisions." We need to learn from Indigenous approaches to governance that prioritize respect and collective responsibility, and apply these millennia-old lessons to twenty-first-century movements for digital democracy.

11

Protect the Digital Commons!

Socialize the Cows

Boston Common is the oldest city park in the United States. It began as a sale of land by William Blackstone, a somewhat eccentric Anglican clergyman, to the city of Boston in 1634. Every townsperson contributed a minimum of six shillings to the purchase. Once the sale was completed, there was some dispute over whether the land should be divided between the inhabitants or held collectively. By a narrow margin, as we will see, the latter plan prevailed, and it has been owned by the city ever since.

Since this decision in the 1630s, Boston Common has been used for a variety of purposes: among others, as a dumping ground, to host the gallows, for sports and games, and as a haven for courting couples. It was home to victory gardens during World War One and donated most of its iron fencing for scrap metal in World War Two. Martin Luther King Jr. spoke to enormous crowds gathered on its lawns, and thousands of people came together there to protest the Vietnam War. It evolved as a community space for all these activities before becoming an official park. "Boston Common belongs to the world," wrote Samuel Barber in his 1914 volume, *Boston Common—A Diary of Notable Events, Incidences, and Neighboring Occurrences*. "From it radiated the influences that led to democratic as opposed to aristocratic rule."

A commons, like its name suggests, is a commonly held resource. It can be something physical or natural like a park or wild environment; or something abstract, like human knowledge about the laws of mathematics, for example. When we talk about the commons, we mean a set of goods or something of value that is not owned by any individual person, though how a common resource is owned and governed may differ. A commons is something that is collectively shared, and its use and protection affect the entire community.

It is almost impossible to talk about the commons without talking about tragedy. In his widely cited essay, *The Tragedy of the Commons* (1968), the ecologist Garrett Hardin explained the problem with cows. If everyone were allowed to let their cows loose on common land, there would be no incentive to stop more and more cows being sent out to graze. Each individual is motivated to reap the greatest possible benefit from this common resource—but the resource is finite. Demand gradually overwhelms supply, as the benefit of adding extra cows flows to the individual cowherd, while the collective cost to the commons does not factor into the calculation. Each person consumes more, at others' expense, until in the end there is no grass for anyone. This is a problem of scarcity, as orthodox economists have been saying for a very long time. The assumption is that humans are not a cooperative species, capable of self-managing a collective resource. Unrestricted demands upon a resource that is finite will inevitably reduce that resource through overexploitation. The commons becomes barren. The commons ends in tragedy.

Hardin's influential essay was not the first to identify this problem; the conundrum and its possible solutions stretch back millennia. The usual response has been to call for the commons to be split up and each cowherd made responsible for looking after their own pasture, policed by some overarching authority. This gives each an interest in using their plot wisely and avoiding its demise through overuse. Put into practice, this is what happened during the "enclosures" of the eighteenth century, a process the historian James Boyle summarizes as the "conversion into private property of something that had formerly been common property." The commons was privatized in response to its overuse. The tragedy of the commons is therefore often used as a justification for private property rights.

Indeed, this was a fate that threatened to befall Boston Common in its early days, as local families let more and more cows loose on the pasture. The breaking up of the commons and distribution of land among residents seemed the most logical step to solve the problem of overgrazing. Yet Boston circumvented tragedy. The town got together to set limits on the amount of permissible grazing and appointed a keeper to monitor compliance. "People often tested the limits, but the system worked through eight generations," writes historian David Fischer in his history of the Common. "It succeeded because the town combined the idea of the common with the institution of the town meeting." This approach to collective management of common resources has been the subject of prolonged research by Nobel Prize–winning economist Elinor Ostrom. She found that tragedy can be avoided by devoting thought and care to the rules for governance of the commons. Ostrom's work has direct relevance to our digital society.

The scholar David Harvey claims to have "lost count" of the number of times Hardin's essay has been cited "as an irrefutable argument for the superior efficiency of private property rights with respect to land and resource uses and, therefore, as an irrefutable justification for privatization." An argument given far less attention by mainstream economists, but which equally avoids the tragedy, is to socialize the cows. (That particular strain of thinking gave rise to *The Communist Manifesto*.) Such reflection ought to encourage us to reexamine the logic of privatization and the supposedly inevitable fate of the commons, as well as the moral justification for the process by which this happened: enclosure.

This period of enclosure was hardly peaceful. It was an act of violence to the fabric of society at the time, described by one historian as "a revolution of the rich against the poor" and "a long, slow, violent operation." This was not some polite rearrangement of resources or a civil attempt to save the poor from themselves. Historians Peter Linebaugh and Marcus Rediker describe how it represented a clash between two classes and was part of broader European colonial projects:

> Since the people of the world have, throughout history, clung stubbornly to the economic independence that comes from possessing their own means of subsistence, whether land or other property, European capitalists had to forcibly expropriate masses of them from their ancestral homelands so that their labor power could be redeployed in new economic projects in new geographic settings.

Enclosure represented a break in traditions of how people carried out their daily activities. It was not a calm, rational transition to a new and functional society. It created a class of politically powerless people, who feature in the historical records as lawless and violent; it was a period of significant riots across England. This tradition of anti-authoritarian rioting against enclosure that began in England found its way to the American colonies. As a form of political resistance to centralized policies like food hoarding, tax collection and military service, riots were a semi-regular event on Boston Common throughout the eighteenth century, variously attributable to groups such as "negroes and persons of vile condition" in 1747 and "the working classes" in 1768. The eradication of subsistence farming ultimately gave rise to the landless working class. The demise of the commons marked a key transition point away from feudalism toward capitalism.

The idea of a commons and its tragedy has largely been associated with natural or physical resources, owned collectively. But the idea of an information commons has existed for almost as long as Boston Common. Since the early seventeenth century, intellectual property rights have been issued over abstract ideas, most commonly over inventions by skilled technicians. Boyle talks about how in its earlier days, intellectual property law actively sought to protect the commons of human knowledge, by granting only limited individual rights over intellectual property, usually for a period of fourteen years. The law understood the importance of a store of common materials for all creators and thinkers to draw from. Intellectual property rights over ideas and knowledge were the exception, not the rule.

But over time, particularly during the twentieth century, there has been a transformation of more of human knowledge into property. Information has become more central to generating value under capitalism in the digital age than it was in the past. We can see this in the contours of various technology platforms and the data mining that happens there. But perhaps more significantly, information has also become a much more valuable commodity in industrial settings. Computerized airplanes, cars and other kinds of productive technology operate not as self-contained pieces of machinery, manufactured, tested and sold, but as nodes in a network constantly relaying information back into the design process. Information, whether in the form of aggregated data, controlled research or network feedback, is an increasingly critical component of our economic system.

As information has become a more valuable resource, intellectual property law has acquired unprecedented clout in modern capitalist relations. Digital rights management over e-books and music, for example, reaches inside our devices and controls how we consume things. Company information, held in the minds of the company's former employees, is locked down by nondisclosure agreements. Takedown notices are issued by courts for peer-to-peer sharing websites. Pharmaceutical companies race to patent various genetic mutations so they can obtain the rights to sell diagnostic testing for that mutation. Through licensing, car manufacturers restrict the kind of diagnostic testing methods that can be used on their vehicle, which is then linked to the kinds of repairs and modifications that can be made. In 2018, a man who built a business helping customers to extend the lifespans of their computers was convicted of conspiracy to traffic in counterfeit goods and criminal copyright infringement. Eric Lundgren had created discs that helped licensed Microsoft users reboot Windows on their old computers when they crashed or needed to be wiped, a perfectly legal reactivation of their license rights. Lundgren was imprisoned for his efforts, with Microsoft assisting prosecutors in the case. The concept of intellectual property law holds such sway in society that it facilitates the imprisonment of people who help other people recycle.

This in turn has influenced how we produce things. Each expansion of property claims over these information resources signals what Boyle calls "a vote of no-confidence in the productive powers of the commons." In other words, it represents an assumption that information will be put to more productive use privately than collectively. As items are withdrawn from the information commons, subjected to private property rights that prevent using and building upon this knowledge, we are all diminished by the shrinkage of the pool of knowledge from which people can generate and develop ideas. Boyle talks about how the increasing spread of intellectual property rights, over everything from software to human cells, is a kind of redistribution in the modern age that is similar to the project of privatizing land centuries ago. In other words, one way to think about this expansion of intellectual property in the digital age is as a second enclosure movement.

Open source software, discussed in chapter 6, gives us some idea of how to arrange production better, and it can even suggest other contexts where this model may be successful. But we need to facilitate collective working environments and protect the fruits of these labors. There must be ways to avoid this work being squirrelled away by companies who want to exploit such common resources for their own gain or to enforce systems of consumption that are egregiously wasteful. We need to socialize the cows.

If privately held information is not to become the norm rather than the exception, we need to consider how we can continue to both generate and guard the commons. We are entering an age when information in various forms, but especially digital forms, amounts to a kind of capital—something that can be directed to the purpose of generating profit, something that affects both what we can learn and how we can produce things. For various reasons, legal, political and moral, the public has a claim over these goods. The extent to which any of this information can or should be in the commons is a question we have not properly begun to grapple with.

Some information ought not be included in the commons, and some information rightly ought to remain in the hands of individuals. Absolute freedom of information is not justifiable, especially given some of the privacy issues discussed in earlier chapters. But we do need a deeper deliberation about the extent to which we should allow information to be privatized and commodified. There is information that causes great harm if kept under lock and key. To what degree are we prepared to tolerate the enclosure of the digital commons to serve the interests of capitalism?

When the townspeople of Boston were deciding how to manage the land they had purchased in 1634, the original plan had been to split it up, and they elected a committee charged with distributing land to residents. One of the members elected was John Winthrop, who was, as it happened, opposed to the idea. Winthrop saw the Common as a way of "joining one person to another and one generation to the next," as Fischer records. His passionate belief was the same as he had expressed upon first arriving in America a few years earlier. "We must be knit together in this work as one man," he had proclaimed then. "We must delight in each other, mourn together, labor and suffer together, always having before our eyes our Commission and Community in the work." He managed to sway others to his view, and the collective decision was made to abandon the plan. The Commons was saved from the fate of enclosure, and Bostonians all enjoy it as a result. It is this collective spirit that must inform the reclamation and protection of the digital commons.

The notion of a digital commons gives us a chance to rethink the hoary old tragedy of the cows. One way to begin is by thinking about what resources ought to be included in digital commons and whether this inclusion creates the same problems as did the proliferating bovines. Creative goods, like books, images, music and software, serve as a good starting point. These goods are unlike almost every other resource that we have used or relied on as a society in the past. This is in part because they are not a scarcity in the economic sense; more specifically, perhaps, they are infinite, in that one person's use of these goods in a digital form does not impair the ability of another person to use them. It makes no difference to the resource how many times it is used, viewed, shared or copied. There is no depletion of *Romeo and Juliet* when an electronic version of the play is shared in the public domain. It is a resource held in the commons, yet no tragedy looms (other than in the storyline).

The classic approach to regulating the production of these works is to impose requirements on users in the form of payment to the author. That makes sense in a book or record store, but it has the potential to be much more than this on the web. Copyright laws increasingly serve to curtail not just access to knowledge generally but also freedom of association and freedom to speak about certain topics. Proposals to crack down on copyright infringements—justified on the basis of the right of authors to proper payment—often give wide-ranging powers to state agencies to filter the web and limit the accessibility of content to those people and platforms that can afford to pay.

We need to think about these goods differently. These goods are non-rivalrous: there is no competition between people to use a particular work, because it is possible to make an infinite number of copies of that work. The cost of producing a copy is zero. As such, it is often difficult to track where these copies go. Every computer connected to the Internet is a kind of printing press, able to copy and distribute copies of a vast amount of human knowledge with just a few clicks. "It is impossible, or at least hard," writes Boyle, "to stop one unit of the good from satisfying an infinite number of users at zero marginal cost." Hence the enormous problem known as piracy. To take an alternative view, then: it is possible to see piracy as the superior mechanism for distributing creative goods.

This is not to say that creative goods do not cost something to create in the first place. Creative works in digital formats take resources to produce or generate the first copy. The writer has to draft her book, just like the geeks in the MIT computer lab had to write the code for the computer game Spacewar. Production of the first copy is a challenge. But the point is that after that initial investment of human time and energy, and some equipment, every single subsequent copy of the creative good can be made at zero cost. The infinitesimal amount of light down a fiber optic cable uploading the file onto a common space is pretty much the only cost of additional copies of these goods.

This is why the digital world of creative goods looks so different to the cows hogging the pasture. The digital commons does not suffer from the problem that the grassy commons had: overuse does not lead to tragedy. In fact, by keeping these goods in the commons, accessible freely, they are distributed most efficiently. That creates a strong argument for holding creative goods in the digital commons.

Are there other goods for which the production of a marginal copy is also zero? Arguably, this is true for certain kinds of information, or goods made up of valuable information. Pharmaceutical companies, for example, carry out research and testing to come up with a specific combination of chemicals to treat a certain condition. The direct cost of producing the drug in question may be relatively low. Yet its price will reflect the inputs of time and research that went into making the first copy, so to speak. The marginal cost of producing a single pill may be close to zero, but pharmaceutical companies pile onto the price the costs of generating the first "copy" of that drug, or the information that went into making that copy, as well as a profit margin.

As with cultural goods, then, it is no surprise that intellectual property law has tightened its grip on production processes in the pharmaceutical industry. Rather than intellectual property serving as a method that generates motivation and drives innovation, it ends up in reverse: the prospect of making a profit determines what is produced. Pharmaceutical companies often go to extraordinary lengths to extend their patents, a practice known as "evergreening," which according to one industry insider necessitates "floors full of lawyers." As Professor Joel Lexchin puts it: "Typically, when you evergreen something, you are not looking at any significant therapeutic advantage. You are looking at a company's economic advantage." Resources are devoted not to improving the function of a drug but to protecting the capacity to profit from that drug.

Software is not that different from pharmaceuticals in this respect. The free software movement talked about "free speech, not free beer" because, not long ago, so-called free software often did have a price—that of disks and postage. But as the Internet facilitated more rapid distribution, the physical cost of free software largely disappeared (though many people still donate money for the programs they use). There are differences when compared to the production of medicines, but the analogy allows us to draw a distinction between what drives the setting of a price and where the value comes from. The valuable part of free software was the code and the hours of time and human ingenuity it represented. With drugs, there is a production cost, but the actual value comes from the time and energy put into the research to arrive at a specific combination of ingredients.

In reality, intellectual property law, in the form of patents and copyright, takes things of value that ought to be in the digital commons and creates a state-sanctioned monopoly to sell them. This makes it harder to access these goods and restricts the extent to which they can be used by others to build upon our collective knowledge. The outcome of this protection is astronomical profits. Companies that are dependent on monopolies created by intellectual property make a lot of money. Tech companies Apple and Microsoft and drug maker Johnson and Johnson are all in the top ten most profitable companies in the Fortune 500 list. On average, pharmaceutical companies enjoy the largest profit margin of the five main industrial sectors, bigger than oil and gas, or banking.

This in turn creates rent-seeking behavior, where the expansion of intellectual property rights becomes a profitable activity for companies that hold these rights. It is not just pharmaceutical companies that spend big on lawyers to protect their patents. Large producers of cultural content, such as Disney and Warner Brothers, have lobbied for extensions of copyright terms to suit their interests as owners, while failing to further either of copyright's core aims, namely to incentivize new cultural production and to reward authors. Software companies make use of open source software to build their products, which they package up and sell. Intellectual property law, therefore, functions as a method of raiding the information commons.

The problem with this is more insidious than it might first appear: it is not just the loss of a body of knowledge that would otherwise be freely available. It depletes the productive power of the commons. The digital commons is the space where projects are worked on, where historical context allows integration with more recent initiatives. It allows people to scour the history of human endeavor to look at how problems have been solved in the past and how this thinking might usefully be applied to modern conundrums. It also facilitates the freedom for people to work in different ways, to spend varying amounts of time and energy on various elements of a particular project. The capacity to work in this way is diminished when the information commons is privatized.

Contributing work to the digital commons is the price we pay for using the resources that are preserved there. This kind of contribution—from each person according to one's ability and skill, to each project according to its need—recalls the aspirations of revolutionaries past. If we are to think about ways in which the kind of open source model seen in the free software movement can be put into practice in other contexts, we need to also understand that it works, in no small part, because there is an abundant and growing digital commons of software to draw from.

The Internet itself demonstrates this in practice. The technical standards that the Internet is based on, and ultimately also the foundations of the World Wide Web, are all part of the digital commons. They are entirely open. Any person can create new programs or features on top of what is already offered in these protocols, without having to ask anyone for permission. As Rebecca MacKinnon writes: "The Internet's inherent value and power come from the fact that it is globally interoperable and decentralized." It might have been otherwise: the basic building blocks of the Internet could have been privatized and licensed out. The closest analogy is cable television (which is a reason why it is commonly used as a metaphor to imagine the world without net neutrality). But unlike cable television, with its restricted access through subscription, we have an Internet that facilitates two-way systems of engagement, where everyone can create their own broadcasting platform for free. Here is an example of the social benefits of the commons post-enclosure. The openness of Internet standards has powered its widespread use; the hosting of its protocols and standards in the digital commons has facilitated this.

The digital commons therefore serves two purposes. First, it facilitates the effective distribution of certain goods. This is especially true when the marginal cost of producing that good is zero. Second, it facilitates production, specifically efficient, collaborative labor. A common body of information avoids duplication and creates economies of scale. It allows open source or peer-to-peer ways of working, the likes of which we have seen deployed so effectively in the free software movement. The idea that the commons can be strip-mined to sacrifice this purpose for the sake of profitability is something that ought to be vigorously challenged. Intellectual property laws hinder both these purposes of the digital commons from being realized.

If we take a step back, then, and think about designing society in a way that makes optimal use of the resources and methods at our disposal, one such test may be: if the marginal copy of a good is zero, it belongs in the digital commons. Setting aside the issue of how we produce the first copy for now, there are good reasons to claim that locating such goods in the commons facilitates superior distribution.

Of course, that does not always mean that such goods must be produced collaboratively. No one would argue, for example, that great works of literature would be improved by designing them in an open source format. *Moby Dick* would hardly be enhanced by crowdsourcing poetic reflections on whaling. There are deep divisions over whether Jeff Buckley or k. d. lang's versions of "Hallelujah" are better than Leonard Cohen's, even though Cohen is the author of the work. Attribution in the age of the digital commons is therefore important. So, while copies can be made and distributed at zero cost, and people should be allowed to work on and with creative works, attribution is central to allowing creative pursuits to be acknowledged. Some singular visions of creativity deserve to be respected as individual pursuits. But that does not mean that we do not all benefit from such works being available for reading and analysis in a commons environment. The world is a smaller place when important human creations are closed off from the public.

However, for non-creative goods, like research output in relation to treating health conditions or safety measures as they operate in autonomous vehicles, for example, collaboration is essential. By standardizing and sharing such information, we can avoid reproducing work (and mistakes) already made in the past. This is a life-or-death matter for many people, not something that ought to be sacrificed to preserve a competitive advantage for proprietary companies. Such information could be held in a public trust, managed by a committee with democratic oversight, with the capacity to set rules as to how access to this information is granted.

In terms of the productive potential of the commons, another test complementary to the one articulated above may be: if information has more value as a common resource than a privately held one, it should be held in the information commons. Data about the safety of a computerized airplane has more value if it is commonly shared within the entire aviation industry rather than privately held. It improves safety and efficiency. Data collected from patients that provides information about the success or failure of medical breakthroughs ought to be publicly available, to allow better and faster progress in research (provided this is done with personal consent and meaningful privacy protections). While personal data needs to be treated with special care, data generated for industrial production and manufacturing purposes has value for us collectively, and its commodification is increasingly hard to justify.

The tragedy of the commons is a lesson of economics that belongs in the twentieth century. Just as enclosure represented the end of feudalism and the beginning of capitalism, the development of the digital commons ought to mark a new beginning in our history of political economy.

If we grow the commons with information and creative goods, the key question becomes how to fund the production of the first copy. The first copy might be a work of art; it might be a drug to cure cancer; it might be the software code for a self-driving car. This question is one of the most important because making the first copy, unlike production of marginal copies, involves a finite resource: human labor. People must devote time and energy to creating goods that can increasingly be shared with ease in the digital age. If we want these goods to build the digital commons, what do we know about what drives these innovations?

As a starting point, we need to discard the idea that intellectual property law is a good system for motivating the production of these first copies. Bill Gates may have lamented the alleged thievery of software by his fellow hobbyists, because it denied him the expected reward for his personal investment. But he was surely the exception, not the rule. The free software movement is a living example of how great slabs of code can be written by volunteers without any promise of profit. There are many reasons that people participate in open source projects, including reputational gain, the desire to solve a problem personally encountered, a sense of achievement, and even sheer joy. One study found that the single most common reason that people participate in open source development projects (accounting for 44 percent of responses) is the pleasure of intellectual stimulation. The old way of thinking, that profit is a key incentive for innovation, has no place in the modern world.

When it comes to creative goods, anyone can appreciate how incentives are not simply monetary. Artists and geeks tend to value other rewards. But a contrast is commonly drawn with informational goods. Working on medical research is arguably more functional than the production of art or open source software, for example. Intellectual property rights are necessary, we are told, to ensure that people are motivated to conduct this kind of research; medical research is hardly as enjoyable as composing music or writing literature or software. But tell that to Jonas Salk, the man who invented the first successful polio vaccine. He was entirely indifferent to the prospect of profiting from his vaccine. When asked who owned the patent on his vaccine, Salk famously replied: "Well, the people, I would say. There is no patent. Could you patent the sun?" In reality, Salk could not have claimed rights associated with his discovery because it was developed using public money. But his views suggest that profitability associated with medical discoveries is not the key motivator for researchers at the highest level in their field. Motivation for human endeavor, whether it is creative, technological or scientific, is more universal in its essence than it might first appear.

There are, of course, people who are solely driven by money. Perhaps surprisingly (for such a profound question of human existence), there is research that claims to be able to quantify this at around 30 percent. At least half, however, "systematically, significantly and predictably behave cooperatively." To paraphrase professor Yochai Benkler, who has written extensively on this topic: we all know people who are teachers, and we all know people who are stockbrokers. So it is a mistake to presume that money is the sole (or even predominant) incentive for human behavior, including in more functional areas of activity. Intellectual property law does not foster development that is inherently efficient or necessarily useful for the purposes of producing the first copy of many kinds of goods.

All this is not to say that people do not need to earn money and feed themselves and their families. This is doubtless one of the main reasons why the left has been hesitant around this argument: there is a fear that if we insist that human labor is about more than its monetary value, it will be assigned a monetary value of zero. How we can satisfy people's material needs and desires is a serious question that warrants careful examination (some suggestions have been offered in earlier chapters). It will involve the radical redistribution of resources, the provision of universal public services, and consideration of how to provide and ensure universal subsistence incomes.

It is possible to imagine subsidies for creatives and others to give them the space to do important work that we all benefit from. But this ought not stop us from rejecting the specious capitalist logic we are sold, which asks us to embrace the idea of intellectual property and surrender to the degradation of the commons.

Creating and fostering a digital commons will not solve all our production problems. Some kinds of production require expertise, and some require specialist equipment. How this can be integrated into open source production processes is an interesting and often complex question. But almost all production involves human time to some degree, and this justifies approaching these questions from the starting point that they ought to be structured as open source, supported by a digital commons.

It is also going to be vital that we think through the implications of commons-based production where it involves personal data. The capacity to collect this kind of information in the digital age creates opportunities to better distribute resources and offers a rich resource for researchers. How we buy food, consume energy or commute en masse, for example, could be analyzed to improve production processes and reduce waste. We can deduce trends from health information collected on a scale larger than ever before possible in human history, offering unprecedented opportunities to advance our understanding of medicine. But it is vital that such data is voluntarily contributed for specific purposes and not sold or reused. We need to respect people's personal sovereignty over their data, so that consenting to it being shared becomes something based on trust and the common good. Moreover, while the benefits of open source production are easily arguable in theory, the implementation has not been without its problems. This is certainly true in the context of software. Some of these problems have stemmed from the inherent nature of open source design, which gives rise to a quasi-obsession with meritocracy. Participants are judged strictly according to their output, with no understanding of how structural realities play into this. For obvious reasons, as we know, almost all the hackers working in the early days of the free software movement were male, and the vast majority of them were white. They almost all hailed from the West and were associated with powerful corporations or universities. The requirement of contributing free labor was a luxury not everyone could afford as easily as they could. It is hardly a surprise to find the inequalities that exist in the world replicated in digital environments. But that does not absolve us of actively making space to identify and overcome them.

These are conundrums that need thinking through in public and transparent ways. But we also need to accept that the digital commons is something worth protecting, which ought to be expanded as much as possible. Next, we need to find ways to keep such information in the commons without compromising our individual privacy. We also then need to work out how to facilitate broad and inclusive participation in open source projects that make use of the commons. All this must be predicated on an understanding that the commons is a vital part of the infrastructure that will allow us to make the most of the digital age, without which we are always going to lose out to the interests of those seeking to profit from the privatization of the commons.

Despite being one of the most visited sites on the Internet and serving as the gateway to all kinds of learning, Wikipedia is often thought of as an anarchic space, lacking in professionalism. It is the butt of many jokes about the unreliability of the Internet, but we lose something if we dismiss it on these terms. The reality is that the site is governed by a complex and evolving set of rules. These rules are broad and flexible but work on the basis that contributors to the platform are able to discuss them and adapt them to suit particular circumstances. Benkler gives one example, involving a dispute about the content of an article about creationism and evolution. The difference in perspective of two contributors to this article was debated and discussed, thrashed out by the participants within the various rules set by Wikipedia that require a neutral point of view and reliable sources. The participants arrived at a compromise without appealing to some higher authority or bureaucratic arbiter. Wikipedia is a space in which the work of contributors is self-governed, and the outcomes of this are reached through transparent means (to the extent that many readers are interested enough to consult the archives).

The website has its flaws. One of the most pressing is what researchers have called "uneven geographies of participation"—the strong overrepresentation of posts about and contributors from the global north. This is a problem of access, but it also represents a preoccupation with particular histories, events and processes, often at the expense of those who exist on the peripheries. But as the researchers concluded, these problems are not insurmountable; rather they demonstrate how "the democratizing potential of Wikipedia has not been realized." Moreover, in spite of its limitations, Wikipedia is more democratic than a book written by a single academic with internal and often unacknowledged bias. This is not to advocate for a rejection of expertise or to dismiss the problem of information gatekeepers being reproduced in digital formats. Rather it is to claim a place for deliberative, informed and transparent methods for assessing human knowledge. Wikipedia has shown how it is possible to manage this.

Wikipedia serves as a working example of how a commonly held resource—in this case a free, collaboratively drafted encyclopedia—can be governed effectively by rules determined in a participatory way. The work of Elinor Ostrom, the Nobel Prize–winning political economist, is instructive in this respect. Ostrom spent her life studying the governance of common resources. She set out a range of guidelines to ensure that these resources would be properly managed, a number of which lend themselves to application in the context of open source production and the digital commons. In particular, her guidelines include ensuring the following: that those affected by the rules can participate in making and modifying the rules; that rules can be matched to specific needs and conditions; that monitoring of behavior is carried out by community members; and that there is an accessible, low-cost method for resolving disputes. These factors are all present in Wikipedia. They provide a good guide for the implementation of digital democracy into the field of information production in the digital age and the management of common resources.

More specifically, this can help us formulate some structural preconditions to the success of open source projects that draw on common resources. Many of these can be found in the glossary for an Agile approach to management, commonly used in the software industry and discussed in chapter 6. They include the capacity for the project to be broken down into smaller parts, for example, so participants can work on pieces that suit their skills and motivation levels (in Agile terms, "user stories"). Benkler also talks about finding ways for people to accept the importance of quality assurances on their work. This may include some double handing (having some tasks done twice) or participants' acceptance of randomized checks. These strategies are commonly used in Agile through pair programming, with workers collaborating on the same computer, and quality assurance tests. It also finds form in pull requests made during software development, where a coder notifies others of changes so they can be checked. Benkler argues that these social norms need to be agreed on, not imposed, for this to work. In other words, as Ostrom suggested, the rules are more likely to be followed if they are drawn up collectively, in less hierarchical formats. People working collaboratively do so better if they understand the rules they are working under and are given a say in how those rules are decided.

GitHub is another worthwhile example to consider. GitHub is an online platform for hosting and managing software as it is developed, so people working on the code can keep track of versions and updates. It gives people a place to host a software project and collaborate easily with others. The concentration and exchange of source code on the platform allows for a different kind of ongoing communication between technologists, industry giants and users. Experimenting with code, submitting pull requests and transparent ratings systems combine to offer the opportunity to create proofs of concept, build reputations, and share and test ideas. Star ratings of code given by peers serves as a form of evidence, a metric of how many forks there are, or how many people are using, copying, building on and changing the code. It is an enormous operation, storing vast amounts of code, a resource powered by rules that are largely defined outside standard market practice. It works because people understand, appreciate and accept the rules.

Projects that draw on the commons and use collaborative working styles can, then, be productive and effective, but not because they are anarchic. They work because they have rules, ones that are quite specific and responsive to the conditions in which these collaborations take place. We need to preserve this "invitational quality" of the commons, as Adam Greenfield puts it, and practices and rules that encourage openness and porosity. "To seal off opportunities for participation is to invite metabolic death," he argues. Instead of relying on a cynical understanding of what motivates people and encouraging them to squirrel away their contributions to compete in a marketplace, we should create structures that encourage freedom to participate. Central to this task is a set of rules for building and using the commons.

GitHub does not exist in a vacuum. It is not immune to the laws of the market. Bought by Microsoft in 2018, its future is unclear. Indeed, the fate of GitHub highlights a significant problem with the commons, namely that under capitalism, it is constantly under threat of being raided. As we know from chapter 6, open source production is capable of creating better products than private enterprise, and having a common repository of code has accelerated production processes in all sorts of ways. But this has not escaped the notice of technology capitalists, who benefit from this resource, contributed by volunteer labor, without having to pay for it. When a common resource exists in a capitalist paradigm, there is inevitably a risk that this resource will be enclosed, absorbed one way or another by the profit motive.

This is a difficult conflict to manage, especially in the absence of an immediate appetite for radical social change. But until we reach that time, the conflict can be an axis upon which to organize and agitate. Just as Ostrom observed the power of participatory rule-making in governing common resources, so too can the defense of the commons be seen as an expanded exercise in generating the social conditions necessary to protect it from the profit motive and determine how it is best put to use. If we socialize the cows, so that they belong to everyone, not only does this protect the pasture from depletion, it also means the milk and cream can't be hoarded by a select few.

When Garry Kasparov lost his game of chess with IBM's supercomputer Deep Blue, he was sanguine about how this came about: "More quantity makes better quality." In other words, the vast computational skills of Deep Blue overcame any qualitative advantages Kasparov may have gained through his experience, creativity and instinct.

It is perhaps ironic that one of the key moves made by Deep Blue was probably the result of a computer bug. The computer, unable to select a move, seemingly chose one at random, which, according to one observer, "sent Garry into a tizzy." It highlights some of the dynamics in the relationship between computers and humans. Humans at times overthink things, whereas computers can make instantaneous decisions based on a sound assessment of what they do and do not know. Computers have no emotions but do have bugs; humans are full of emotions, and these can be their downfall. Computers have an immense and growing capacity to crunch information well beyond any human brain but are ultimately dependent on input from humans. Computers are impressive and useful; humans have other skills that are equally indispensable in solving problems. Some tasks are evidently done better by humans, others by computers.

Computers work best when there is good data, properly obtained, and collectively shared—when there is a respected and flourishing commons. The commons is the foundation for forms of production that make the most of human talent. Eben Moglen calls this "Moglen's Metaphorical Corollary to Faraday's Law":

> If you wrap the Internet around every person on the planet and spin the planet, software flows in the network. It's an emergent property of connected human minds that they create things for one another's pleasure and to conquer their uneasy sense of being too alone.

Society has arrived at the point where productive relations are acting as an impediment to the development of productive forces. The old style of doing business—making money from proprietary software, packaging up information as a commodity to preserve its value—is slowing us down as a species. As Nick Srnicek and Alex Williams explain, capitalism "misattributes the sources of technological development, places creativity in a straitjacket of capitalist accumulation, constrains the social imagination within the parameters of cost–benefit analyses and attacks profit-destroying innovations." The state plays a role also: if such innovations are not commodified by private enterprise, they are captured and controlled by state apparatuses for their own purposes. This orients us toward technologies of oppression and removes scientific knowledge from the public domain, enclosing it in the classified world.

Digital technology in combination with the information commons gives us the chance to accelerate thinking, learning and problem solving. I envision a digital space where we are all encouraged to take collective responsibility, and we all have ownership of both the pasture and the cows.

Conclusion

History Is for the Future

Another World Is on Its Way

"There's a time when the operation of the machine becomes so odious, makes you so sick at heart, that you can't take part," declared the student radical Mario Savio in 1964 on the steps of Sproul Hall, at the University of California at Berkeley:

> And you've got to put your bodies upon the gears and upon the wheels … upon the levers, upon all the apparatus, and you've got to make it stop. And you've got to indicate to the people who run it, to the people who own it, that unless you're free, the machine will be prevented from working at all.

We are facing a future in which some of the best technological developments are made in relation to warfare or commerce rather than freedom and empowerment. Digital technology has become a machinery for producing billionaires rather than lives of dignity for the billions. While many of the decisions made by the elite of technology capitalism are made in California, the effects are felt all around the world, from the factories of Shenzhen to the mines of the Democratic Republic of the Congo to the streets of London and the train stations of New Delhi. The machinery of the digital age is designed not by us but for us—which is a source of oppression but also a source of power.

Digital technology, for all its scope to watch us, think for us and automate our jobs, is still reliant on us to operate. It is not a mysterious or unexplained apparatus; it creates places that both shape us and can be shaped by us; it is created by people, it is guided by decisions made by people, it is owned by people. People can also stop it from working. We have the opportunity to reclaim the power of technology, to appropriate the machinery and use its gears, wheels and levers, its silicon and glass, and repurpose it for the good of the many rather than the few.

Devising a usable past for digital technology endows it with tradition in an age dominated by ephemera, provides a set of theories for understanding confected realities, a history for a world obsessively chasing the future. It creates a possibility of common cause for an atomized society.

We need yardsticks to assess technological developments and comparisons that help put them in historical context. As technology enters more of the private spaces where we define our sense of self, we should ask whether such incursions are manipulative or facilitate digital self-determination. We should consider how advances in technology do or do not redistribute wealth and the extent to which this allows for the creation of a commons to drive further innovation. As we continue to build the technological and legal infrastructure of our online lives, we should ensure that there are ways for deliberating these developments collectively, for navigating how we protect autonomy, while also maintaining accountability. Our progress should be measured by our capacity to protect our digital environment, and we must think outside of traditional models of governance to achieve this. We should be nurturing collaborative forms of production and the unfettered use of our collective knowledge. Like the radical thinkers and activists that populate our usable past, I want us to challenge the centralization of power under capitalism toward the ruling class and their backers in government.

Technology alone will never solve our problems. But political activism, informed by theory and history, can push for technological development to serve people rather than profit. Digital activists cannot carry out this work alone. They need to work collaboratively with those struggling for justice in other fields. We need them in social movements and organizations, working with fellow activists to design technologies and laws based on radical and democratic principles. Participants in social movements across the spectrum of progressive politics cannot be expected to understand the minutiae of technological design agendas, but they should know who to ask. They must engage with the more general political conundrums posed by digital technology, because we all have a stake in the outcome. The people who make technology need feedback from users so their work can be informed by a broad range of human experiences and serve the public effectively. Together we can mobilize society's technological sophistication to dismantle capitalism and recreate representative structures that are genuinely democratic. We have the capacity to create communal luxury, but we will have to work collaboratively to make it happen.

Our priority must be to build bridges between people with interests in technology, politics, history, data science, art and activism. There is no denying that grave challenges lie ahead. But it is possible to imagine a future defined by fairness, empowerment and collective joy. Harnessing the promise of digital technology for a public commons makes this more possible than ever. It will be a future of beautiful automatons, curious flâneurs, robots that eat the rich, and socialized cows—wandering freely in our imaginations and our cyberspace.

Acknowledgments

This book would not be possible without the input of many people, of course. We all worked on it together. Thanks for letting me rope you all in, and hopefully I've not embarrassed you (well, not here at least). If there are errors, they are mine absolutely, but, like so many instances of authorship, this remains a collective effort.

I am deeply grateful to my editor, Jessie Kindig, who stuck with me and devoted many hours to helping this manuscript come into being. She is a thoughtful and generous critic and regularly saw things in my work that I had not yet seen myself, which is a wonderful skill. I am deeply indebted to her for that. I am also grateful to Anthony Arnove, for taking a risk on me, and Roisin Davis for helping me get through it, with kindness and encouragement. Thank you to Leo Hollis for his support and openness to my ideas. Thank you to Antony Loewenstein for his advice and encouragement through various stages of writing and publication. Thank you to Lorna Scott Fox for patiently, painstaking editing my words—this book is so much the better for it. Thank you to Mark Martin for his patience and thoughtfulness. Editors are so rarely given their due—too often only noticed when their work is not done well. Here I feel they have done quite the opposite and want to acknowledge them doubly so because of it.

Eben Moglen, you got me started on all this. I hold you legally responsible.

My readers of versions of this manuscript include the indomitable Felicity Ruby, who has always astonished me with her abundant enthusiasm and endless spirit of generosity. Thank you to Keith Dodds and Alex Kelly for your helpful and intelligent comments. Amy McQuire, thank you, all power to you. Jane Brophy, you are so very smart and particularly talented at sharing that intelligence in uplifting ways. I am so grateful for all the time and positive energy you gave me so selflessly! Scott Ludlam, your encouragement was gratefully received and your kind comments deeply valued. Jacinda Woodhead, you are a remarkable editor and woman, fiercely intelligent, always kind, never daunted. Thank you, Joe Shaw, for your time when you had so little to spare. Thank you, Rebecca Giblin and François Petitjean, for your kindness in giving it a pass on the sniff test. Thank you all, dear readers, I owe a lot to you, and you deserve far more credit than you will ever be given for the wonderful work you did to get this thing into shape.

I am lucky to have so many wonderful, intelligent and generous friends, to whom I owe much. I put Alexa O'Brien in that camp, who always showed great faith in my abilities and has routinely demonstrated inspiring bravery. Liz Humphries, Kathleen MacLeod, Michael Brull, Brami Jegan, Scott Cosgriff and Katie Robertson all offered me invaluable assistance, often without realizing how much, I am sure. Jacob Varghese, you have always given a lot of time to exploring endless arguments about tiny differences. Brooke Dellavedova, you've always been top-notch to work for and taught me so much about getting stuff done with grace. Kimi Nishimura, you're very impressive, and I'm so pleased to have you around. Kelly Matheson, thanks for looking after me and being a leader with integrity and compassion. I am fortunate to have friends like Nicole Papaleo, Michael Stevens, Justin Peysack, Jeff Sparrow, Guy Rundle, Tim Singleton Norton, Tom Sulston, George Newhouse, Arundhati Katju, Annie Mulroy, Gulika Reddy, Mary Kostakidis, Ian Wilcox, Pauline Spencer, David Yarrow, Ying Qian, David Brophy, Charles Livingstone, Angela Rintoul, Asher Wolf, Leanne O'Donnell, Suelette Dreyfus, Peter Fitzgerald, Giordano Nanni,

Meribah Rose, and Benjamin Laird—who all at various points listened to me, inspired me, encouraged me, turned a blind eye to my shortcomings, or just gave terrific chat. Thanks to all my lovely colleagues and friends at Maurice Blackburn Lawyers, who do their best every day to live up to the name of that great man and his formidable wife, Doris. Paul Bendat, I am so sorry that you're not around to read this and hear me tell you how much I owe to you. Mr. and Mrs. Grace Gotham (David and Sarah Liston): spending time with you while I was writing this always made me feel like a million bucks. To all my pals online who probably don't even realize how important they have been to this project, thanks for sharing fascinating content on the web, keep doing it. All of you: I am deeply touched by how encouraging you have been of me, especially at times when I felt below the task at hand. I have such a ripper group of friends!

I love the British Library, the Battersea Park Library, the New York Public Library, the State Library of Victoria, various unauthorized online libraries and just libraries everywhere. Working in libraries is wonderful. Where would we be without them? Librarians are some of the most important people in the world.

I'm grateful to my clever, giving, witty, never-dull parents, Anne and Bill, for great chats and excellent hospitality. I also owe much to my two wonderful sisters, Katherine and Louise, who were always up for a debate, always quick to be loyal and reassuring. And thank you to their lovely partners, Ralph and Steph, and their gorgeous kids, Martha, Ben and Jean. Thanks Tony Randle, Ashlea Randle and Amit Maini for all the love. In fact, all of my family are pretty excellent, the O'Sheas, the O'Briens, the Randles and the Raos. I miss my grandfather, with his sharp intellect and love of language. I miss my beloved grandmother, who though she made it to ninety-nine years of age, was never tired of life; though she was not given the opportunity to finish school, was deeply in love with learning.

I owe the most to my wonderful partner, Justin Randle. He is my number one fan, and it is a privilege to have someone like him in your corner. He gives me confidence when I am my own worst enemy, he defends me relentlessly even from myself. He makes me strive to be the best version of myself, and he never ever loses sight of that person. He was pretty much the only person who laughed when I made a joke about Margaret Thatcher's death in front of a few thousand members of the legal profession (most of them booed). If I were going into battle with the worst collection of capitalist vampire squids and cyborg spooks, I would want him on my side. Muhammad Ali said, "it ain't bragging if you can back it up." I ain't bragging: he's the greatest.

Notes

p. 1, **violent nature:** Henry Kamen, *Philip of Spain*, Yale University Press, 1998, 120.

p. 2, **possibly on an illicit errand:** Andrew Villalon, "Putting Don Carlos Together Again: Treatment of a Head Injury in Sixteenth-Century Spain," *Sixteenth Century Journal* 26:2, 1995, 350; Elizabeth King, "Clockwork Prayer: A Sixteenth-Century Mechanical Monk," *Blackbird*, Spring 2002.

p. 2, **hole in his skull:** Villalon, "Putting Don Carlos Together Again," 355.

p. 2, **desiccated corpse:** Ibid., 356.

p. 2, **he would recover:** Ibid., 361–2.

p. 2, **made of wood and iron:** King, "Clockwork Prayer," 1.

p. 3, **A prodigy from humble origins:** Ibid., 18.

p. 3, **ambitious impulse … and artificial intelligence:** Ibid., 47.

p. 4, **Would the measure of the monk's power:** Ibid., 61.

p. 5, **The cure was of natural origins:** Villalon, "Putting Don Carlos Together Again," 363.

p. 6, **a person profoundly committed:** James R. Vitelli, *Van Wyck Brooks*, Twayne, 1969, Preface, i.

p. 7, **The present is a void:** Van Wyck Brooks, "On Creating a Usable Past," *Dial*, April 11, 1918, 337.

p. 7, **devoted years of his life:** Raymond Nelson, *Van Wyck Brooks: A Writer's Life*, E. P. Dutton, 1981, 210.

p. 7, **cultural centralization:** Vitelli, *Van Wyck Brooks*, 36.

p. 8, **All our invention:** Karl Marx, *Speech at Anniversary of the* People's Paper, 1856, marxists.org.

p. 9, **When you give everyone a voice:** Saroj Kar, "The Best of Mark Zuckerberg," *SiliconANGLE*, May 23, 2012, siliconangle.com.

p. 9, **great entrepreneurial CEOs:** John Patrick Leary, "The Poverty of Entrepreneurship: The Silicon Valley Theory of History," *New Inquiry*, June 9, 2017, thenewinquiry.com.

p. 9, **Knowing that others:** Van Wyck Brooks, "On Creating a Usable Past," *Dial*, April 11, 1918, 341.

p. 11, **What would life be:** Letter from Vincent van Gogh to his brother Theo, The Hague, December 29, 1881, webexhibits.org/vangogh/letter.

p. 12, **a frustrated archaeologist:** Corinne Maier, *Freud: An Illustrated Biography*, Nobrow, 2013, 22.

p. 14, **if you're not paranoid:** Walter Kirn, "If You're Not Paranoid, You're Crazy," *Atlantic*, November 2015, theatlantic.com.

p. 15, **The idea of privacy:** See Colin Bennett, *The Privacy Advocates*, MIT Press, 2008, 3–6, for a nuanced discussion of the various meanings of privacy.

p. 17, **cyber vampire squids:** A metaphor adapted from Matt Taibbi, "The Great American Bubble Machine," *Rolling Stone*, April 5, 2010, rolling stone.com.

p. 17, **Target Stores predicted a teen was pregnant:** Charles Duhigg, "How Companies Learn Your Secrets," *New York Times Magazine*, February 16, 2012, nytimes.com.

p. 17, **We do that for grocery products:** Ibid.

p. 17, **to identify those unique moments:** Ibid.

p. 18, **It can be hard to obtain:** Katy Bachman, "Big Data Added $156 Billion in Revenue to Economy Last Year [Updated]," *AdWeek*, October 14, 2013, adweek.com.

p. 18, **information can then be matched to other data sets:** Casey Johnston, "Data Brokers Won't Even Tell the Government How It Uses, Sells Your Data," *Ars Technica*, December 21, 2013, arstechnica.com.

p. 18, **data on smartphones is often siloed:** See Antonio García Martínez, *Chaos Monkeys*, Harper Collins, 2016, "The Great Awakening."

p. 18, **a more customized approach:** Richard L. Tso, "Retail's Next Big Bet: iBeacon Promise Geolocation Technologies," *Wired*, May 2014, wired.com.

p. 19, **another way to know a user's whereabouts:** Kim Komando, "These 7 Apps Are among the Worst at Protecting Privacy," *USA Today*, September 18, 2015, usatoday.com.

p. 19, **digital Trojan horses:** Researchers from Carnegie Mellon University have a project called Privacy Grade, which assigns letter grades to apps based on certain criteria. One criterion is whether users expect the app to be harvesting data. This means apps that provide maps score better, because users know that their locational data is being used. But many popular free games and tools, most infamously smartphone flashlight apps, score surprisingly badly. See privacygrade.org.

p. 19, **This can track what ads a person sees:** Dan Goodin, "Beware of Ads that Use Inaudible Sound to Link Your Phone, TV, tablet, and PC," *Ars Technica*, November 13, 2013, arstechnica.com. See also Lily Hay Newman, "How to Block the Ultrasonic Signals You Didn't Know Were Tracking You," *Wired*, November 3, 2016, wired.com.

p. 19, **something the advertising industry is highly enthusiastic about:** The American Association of Advertising Agencies states: "Geo-based addressable data driven TV can offer great efficiencies in delivery by using [video] subscriber data to identify income, ethnicity, pet ownership and even purchasing behavior within specific geographies on the households or cable zones level that overindex against the desired target," AAAA, *Data Driven Video: What Will It Mean to the Future of Video*, 2015, aaaa.org.

p. 19, **constantly observing what we do:** See Adam Greenfield, *Radical Technologies*, Verso, 2017, 27.

p. 19, **has effectively erased any privacy safeguards:** Center for Digital Democracy, *Big Data Is Watching: Growing Digital Data Surveillance of Consumers by ISPs and Other Leading Video Providers*, March 2016. See also Cracked Labs, *Corporate Surveillance in Everyday Life*, June 2017, crackedlabs.org.

p. 20, **over 99 percent accuracy:** Yinzhi Cao, Song Li Lehigh, Erik Wijmans, *(Cross-)Browser Fingerprinting via OS and Hardware Level Features*, yinzhicao.org.

p. 20, **The industry is also adopting various forms of biometric profiling:** See Cracked Labs, *Corporate Surveillance*.

p. 20, **you might be identified even when using an anonymized browser:** This applies to users of Tor. Dan Goodin, "How the Way You Type Can Shatter Anonymity—Even on Tor," *Ars Technica*, July 28, 2015, arstechnica.com.

p. 20, **Facial recognition software is already prevalent in the retail industry:** Chris Frey, "Revealed: How Facial Recognition Has Invaded Shops—and Your Privacy," *Guardian*, March 3, 2016, theguardian.com; Max Nisen and Leo Mirani, "The Nine Companies That Know More about You than Google or Facebook," *artz*, May 27, 2014, qz.com.

p. 20, **Shoshana Zuboff has called "surveillance capitalism":** Shoshana Zuboff, "Big other: surveillance capitalism and the prospects of an information civilization," *Journal of Information Technology*, 30:1, March 2015, 75–89. "'Big data' is above all the foundational component in a deeply intentional and highly consequential new logic of accumulation that I call *surveillance capitalism*. This new form of information capitalism aims to predict and modify human behavior as a means to produce revenue and market control."

p. 20, **If iron ore was the raw material:** See stopdatamining.me.

p. 21, **This is merely a "don't-look-too-closely" claim:** Center for Digital Democracy, *Big Data Is Watching*.

p. 21, **87 percent of Americans could be uniquely identified:** Latanya Sweeney, "Simple Demographics Often Identify People Uniquely," Carnegie Mellon University, Data Privacy Working Paper 3, 2000, 2.

p. 22, **the choices it makes about you:** See Greenfield, *Radical Technologies*, 232–3.

p. 22, **like zombies:** See Bernard E. Harcourt, *Exposed: Desire and Disobedience in the Digital Age*, Harvard University Press, 2016. He calls them "digital doppelgangers."

p. 22, **Privacy is the right to a self:** Yohana Desta, "Read Edward Snowden's Moving Speech about Why Privacy Is Something to Protect," *Vanity Fair*, September 16, 2016, vanityfair.com.

p. 23, **production of pleasure:** Sigmund Freud, "Beyond the Pleasure Principle," in the *Standard Edition of the Complete Psychological Works of Sigmund Freud*, vol. 8, Hogarth Press, 1962, 7.

p. 23, **to work over in the mind:** Ibid., 16, see also 17, 35.

p. 24, **precisely between the two:** Mladen Dolar, "Freud and the Political," *Unbound*, 4, 2008, 25.

pp. 24–5, **I will define technology capitalism:** See Luis Suarez-Villa, *Technocapitalism*, Temple University Press, 2009. Suarez-Villa's *technocapitalism* is a form of capitalism, grounded in corporate power, which places a particular focus on the exploitation of technological creativity. My idea is more specific: I am trying to demarcate a section of the industry that has a mission to organize production and social activities around digital technology.

p. 25, **My aim is to use the term to demarcate:** See Richard Seymour, "Ubercapitalism and the Trillion Dollar Reward," December 27, 2017. "They streamline capitalism by *reorganizing it around the format of the computer*," patreon.com/posts.

p. 25, **Amazon has a record:** Adam Tanner, "Different Customers, Different Prices, Thanks to Big Data," *Forbes*, April 14, 2014, forbes.com.

p. 25, **Google has a log:** Becca Caddy, "Google Tracks Everything You Do: Here's How to Delete It," *Wired*, January 20, 2017, wired.co.uk. After acquiring Nest in 2014, which started out as a smart thermostat, Google now also collects data about when you are at home. See Casey Johnston, "What Google Can Really Do with Nest, or Really, Nest's Data," *Ars Technica*, January 15, 2014, arstechnica.com.

p. 25, **These buttons are actually:** Tom Simonite, "Facebook's Like Buttons Will Soon Track Your Web Browsing to Target Ads," *MIT Technology Review*, September 16, 2015, technologyreview.com.

p. 25, **feed data back to Google:** Cracked Labs, *Corporate Surveillance*.

p. 25, **These platforms are the places:** Note that Facebook adds to this by buying data about its users, including mortgage histories and other shopping habits, from data brokers. See Julia Angwin, Terry Parris Jr. and Surya Mattu, "What Facebook Knows about You," ProPublica, September 28, 2016, propublica.org.

p. 25, **A high-fidelity, persistent, and immutable pseudonym:** Martínez, *Chaos Monkeys*, "The Great Awakening."

p. 26, **everything from people:** Angwin, Parris and Mattu, "What Facebook Knows about You." ProPublica created a tool you can use to get a glimpse of what Facebook thinks you like, which collates the inferences it has drawn based on the data it has collected about you. See google.com/webstore/detail/what-facebook-thinks.

p. 26, **no two people's experience of the web:** Joseph Turow, *Niche Envy: Marketing Discrimination in the Digital Age*, MIT Press, 2008, 2.

p. 26, **Peddlers evaluated you … back and forth:** Kaveh Waddell, "Incessant Consumer Surveillance Is Leaking into Physical Stores," *Atlantic*, October 20, 2016, theatlantic.com.

p. 26, **The rich see a different Internet:** Michael Fertik, "The Rich See a Different Internet Than the Poor," *Scientific American*, February 1, 2013, scientificamerican.com.

p. 27, **the empty chair:** George Anders, "Inside Amazon's Idea Machine: How Bezos Decodes Customers," *Forbes*, April 4, 2012, forbes.com.

p. 27, **Acxiom, has encouraged people:** Natasha Singer, "A Data Broker Offers a Peek Behind the Curtain," *New York Times*, August 31, 2013, nytimes.com.

p. 28, **The goal is to distribute:** Harcourt, *Exposed*, 94.

p. 28, **If you consistently get ads:** Joseph Turow, *The Daily You: How the New Advertising Industry Is Defining Your Identity and Your Worth*, Yale University Press, 2011, Introduction. See also Turow, *Niche Envy*.

p. 28, **reputation silos:** Ibid., Introduction.

p. 28, **no single collective experience online:** Julia Angwin, Terry Parris Jr. and Surya Mattu, "When Algorithms Decide What You Pay," ProPublica, October 5, 2016, propublica.org.

p. 29, **By emphasizing the individual:** Turow, *Niche Envy*, 3.

p. 29, **what makes a city center magnetic:** Jane Jacobs, "Downtown Is for People," *Fortune Magazine*, 1958, fortune.com.

p. 29, **intricate and close-grained diversity:** Jane Jacobs, *The Death and Life of Great American Cities*, Vintage, 1992, 39.

p. 30, **Jeremy Bentham's nineteenth-century model:** See Harcourt, *Exposed*, passim.

p. 30, **The watching works best:** Ibid., 127.

p. 31, **Dating websites are a good example:** The process of using the dating site OkCupid is examined in Mara Einstein, *Black Ops Advertising*, OR Books, 2016, 146–7.

p. 31, **The company has admitted … That's how websites work:** Christian Rudder, "We Experiment on Human Beings!" July 27, 2014, theblog.okcupid.com.

p. 31, **The money is in the data:** Einstein, *Blackops Advertising*, 146–7. Match Group underwent a share offering in late 2015 with a valuation of $2.9 billion. Leslie Picker and Mike Isaac, "For Its I.P.O., Square Scales Back Valuation by $3 Billion," *New York Times*, November 18, 2015, nytimes.com.

p. 31, **OkCupid's parent company, Match Group:** Match Group Business Overview, November 2016, ir.mtch.com.

p. 31, **Years of remote controls … clandestine marketing:** Einstein, *Blackops Advertising*, 12.

p. 32, **The programme of becoming happy … all that we desire:** Sigmund Freud, *Civilization and Its Discontents*, Hogarth Press, 1930 (trans. James Strachey), 14.

p. 32, **There are many troubling parallels:** This is a comparison that has also been drawn by those studying addiction and compulsion in online behaviors. See Elias Aboujaoude, Vladan Starcevic, "The Rise of Online Impulsivity: A Public Health Issue," 3, November 2016.

p. 32, **effective marketing messages:** Ibid., 151. Software designers trumpet the fact that casinos increased their revenues by a fifth in as little as eight months using their product

p. 32, **trapping and ultimately annihilating:** Natasha Schüll, *Addiction by Design*, Princeton University Press, 2011, 179.

p. 32, **Venue operators regularly use real-time monitoring:** Ibid., 145, 149, 152, 154.

p. 33, **the painstaking efforts:** Ibid., 56.

p. 33, **the dynamic interaction between the two:** Ibid., 20. In the context of Schüll's observation, I note that the research around the nature of addiction and how it functions—to an outside observer at least—appears to be constantly evolving. As more work is done in this area, we may better understand how this plays out in digital contexts.

p. 33, **exploited every piece of psychological gimcrackery:** Martínez, *Chaos Monkeys*, "Monetizing the Tumor."

p. 33, **We become reliant on access:** Greenfield, *Radical Technologies*, 22.

p. 34, **features that provide intermittent rewards:** Tristan Harris, "How Technology Hijacks People's Minds—from a Magician and Google's Design Ethicist," May 18, 2016, medium.com.

p. 34, **the necessary critique:** See Audrey Watters, "The Tech 'Regrets' Industry," February 16, 2018, audreywatters.com; and Jacob Silverman, "The Techies Who Said Sorry," *Baffler*, March 20, 2018, baffler.com.

p. 34, **the digital equivalent of the perfect hallucinogen:** Harcourt, *Exposed*, 45–6.

p. 34, **Facebook was legalized crack:** Martínez, *Chaos Monkeys*, "The Various Futures of the Forking Paths."

p. 34, **As Neil Postman argued back in 1985:** Neil Postman, *Amusing Ourselves to Death: Public Discourse in the Age of Show Business*, Penguin, 2006, 36.

p. 34, **Orwell's prophecies are of small relevance:** Ibid., 281.

p. 35, **It is very discouraging:** Anthony Paletta, "Story of Cities #32: Jane Jacobs v Robert Moses, Battle of New York's Urban Titans," *Guardian*, April 28, 2016, theguardian.com.

p. 36, **The best minds of my generation:** Ashlee Vance, "This Tech Bubble Is Different," *Bloomberg*, April 14, 2011, bloomberg.com.

p. 36, **to take 1 million cars:** "Taking 1 Million Cars Off the Road in New York City," Uber News Room, July 10, 2015, newsroom.uber.com.

p. 36, **Average travel speeds:** Lauria Bliss, "How to Fix New York City's 'Unsustainable' Traffic Woes," CityLab, December 21, 2017, citylab.com.

p. 36, **Revenue for the city has also fallen:** Emma G. Fitzsimmons and Winnie Hu, "The Downside of Ride-Hailing: More New York City Gridlock," *New York Times*, March 6, 2017, nytimes.com.

pp. 36–7, **London fares no better:** Sam Knight, "How Uber Conquered London," *Guardian*, April 27, 2016, theguardian.com.

p. 37, **reversing much of the environmental and practical gains:** Tanya Powley, Madhumita Murgia, Robert Wright and Leslie Hook, "How Uber and London Ended Up in a Taxi War," *Financial Times*, September 28, 2017, ft.com.

p. 37, **She felt that cities were best understood:** See Jacobs, "Downtown Is for People," and Jacobs, *The Death and Life of Great American Cities*, 755.

p. 37, **To be away from home:** Charles Baudelaire, *The Painter of Modern Life and Other Essays*, trans. Jonathan Mayne, 9. "The artist, man of the world, man of the crowd, and child," columbia.edu/itc.

p. 38, **every downtown can capitalize ... rebuilding a living one takes imagination:** Jacobs, "Downtown Is for People."

p. 38, **participates fully through observation:** Bijan Stephen, "In Praise of the Flâneur," *Paris Review*, October 17, 2013, theparisreview.org.

p. 39, **the West Indian Merchants pooled their funds:** Peter Linebaugh, *The London Hanged: Crime and Civil Society in the Eighteenth Century*, Verso, 2003, 425.

p. 40, **a kind of "collective bargaining":** Ibid., 406.

p. 40, **He teamed up with Patrick Colquhoun:** See John Harriott, *Struggles through Life*, vol. 2, C. and W. Galabin, 1807, 335–7; Linebaugh, *The London Hanged*, 427. Linebaugh describes Colquhoun in the following terms: "If a single individual could be said to be the planner and theorist of class struggle in the metropolis it would be he."

p. 41, **They enforced working hours:** Linebaugh, *The London Hanged*, 433.

p. 41, **bringing into reasonable order:** Harriott, *Struggles through Life*, 337.

p. 41, **he saw the urgency of promoting:** Linebaugh, *The London Hanged*, 427.

p. 41, **Police in this country ... the attention of all:** Patrick Colquhoun, *A Treatise on the Police of the Metropolis*, 7th ed., Bye and Law, 1806, Preface.

p. 41, **It is the dread ... the commission of offences:** Patrick Colquhoun, *A Treatise on the Commerce and Police of the River Thames*, 6th ed., Baldwin and Sons, 1800, 265–6.

p. 42, **Civil government ... who have none at all:** Adam Smith, *An Inquiry into the Nature and Causes of the Wealth of Nations*, Lincoln and Gleason, 1804, 169, 172.

p. 42, **to create crime:** Linebaugh, *The London Hanged*, 434–5. Linebaugh writes: "The historic function of the River Thames Police was less the apprehension of criminals or the detection of crime—and this was well understood by Colquhoun, whose emphasis was always on the *prevention* of crime and the *renovation* of morals—than it was, paradoxical as it may appear, the creation of crime or the practical redefining of its meaning … to create, and then to sustain the class relations in the production of private property."

p. 42, **the Marine Police Office:** Ibid., 426.

p. 42, **Peel himself had learnt the importance:** Alex Vitale, *The End of Policing*, Verso, 2017, 35.

p. 42, **social and political havoc:** Ibid., 36.

p. 43, **The origins and functions:** Ibid., 27.

p. 44, **the relatively tiny sum:** Chris Taylor, "Through a PRISM, Darkly: Tech World's $20 Million Nightmare," *Mashable*, June 7, 2013, mashable.com.

p. 44, **20 billion communication events:** Glenn Greenwald, *No Place to Hide: Edward Snowden, the NSA and the Surveillance State*, Penguin, 2014, "Collect It all."

p. 44, **Collect it All"** Ibid. Janus Kopfstein, "The NSA Can 'Collect-It-All,' But What Will It Do with Our Data Next?" *Daily Beast*, April 16, 2014, dailybeast.com.

p. 44, **National Interests, Money and Egos:** Greenwald, *No Place to Hide*, "Collect It all."

p. 44, **maintain its grip on the world:** Ibid.

p. 44, **a committee for managing:** Karl Marx and Friedrich Engels, *Manifesto of the Communist Party*, 1848, chapter 1, marxists.org.

p. 44, **Multiple studies confirm:** For a list of these studies, see Karen Gullo, "Surveillance Chills Speech—as New Studies Show—and Free Association Suffers," EFF Blog, May 19, 2016, eff.org; Elizabeth Stoycheff, G. Scott Burgess and Maria Clara Martucci, "Online Censorship and Digital Surveillance: The Relationship between Suppression Technologies and Democratization across Countries," *Information, Communication and Society*, September 14, 2018.

p. 45, **Paul Ohm talks about:** Paul Ohm, "Don't Build a Database of Ruin," *Harvard Business Review*, August 23, 2012.

p. 45, **British police have paid:** Sidney Fussell, "British Cops Bought Marketing Data to Help Profile Criminal Suspects," *Gizmodo*, April 14, 2018, gizmodo.com.

p. 45, **American police have used:** Rebecca Robbins, "The Golden State Killer Case Was Cracked with a Genealogy Website," *Scientific American*, April 28, 2018, scientificamerican.com.

p. 45, **It can be hard to know:** Clara Jeffery and Monika Bauerlein, "Where Does Facebook Stop and the NSA Begin?," *Mother Jones*, November–December 2013, motherjones.com.

p. 46, **"Police Cloud" is a public program … unique national identification numbers:** Human Rights Watch, "China: Police 'Big Data' Systems

Violate Privacy, Target Dissent," November 19, 2017, hrw.org.

p. 46, **Researchers are developing:** Josh Chin and Liza Lin, "China's All-Seeing Surveillance State Is Reading Its Citizens' Faces," *Straits Times*, July 8, 2017, straitstimes.com.

p. 46, **A jaywalker caught:** Ibid.

p. 46, **In 2014, the Chinese state:** Anna Mitchell and Larry Diamond, "China's Surveillance State Should Scare Everyone," *Atlantic*, February 2, 2018, atlantic.com.

pp. 46–7, **sensitive to particular purchases:** Mitchell and Diamond, "China's Surveillance State"; Scott Cendrowski, "Here Are the Companies That Could Join China's Orwellian Behavior Grading Scheme," *Fortune*, November 29, 2016, fortune.com.

p. 47, **A citizen's score affects them in myriad ways ... Those who score poorly:** Mara Hvistendahl, "Inside China's Vast New Experiment in Social Ranking," *Wired*, December 14, 2017, wired.com.

p. 47 **an important basis:** "Planning Outline for the Construction of a Social Credit System (2014–2020)," *China Copyright and Media*, June 14, 2017, chinacopyrightandmedia.wordpress.com.

p. 48, **In contrast, the American approach:** Kopfstein, "The NSA Can 'Collect-it-All'."

p. 48, **Numerous profitable companies:** See for example, Sam Biddle and Spencer Woodman, "These Are the Technology Firms Lining Up to Build Trump's Extreme Vetting Program," *Intercept*, August 7, 2017, theintercept.com.

p. 48, **All data is credit data:** Astra Taylor and Jathan Sadowski, "How Companies Turn Your Facebook Activity Into a Credit Score," *Nation*, May 27, 2015, nation.com; see also zestfinance.com.

p. 48, **data mining and analysis technology:** Yasha Levine, "Surveillance Valley," *Baffler*, February 6, 2018; Kate Conger, "Google Is Helping the Pentagon Build AI for Drones," *Gizmodo*, March 7, 2018, gizmodo.com.

p. 48, **Amazon is working:** Ava Kofman, "Amazon Partnership with British Police Alarms Privacy Advocates," *The Intercept*, March 9, 2018, theintercept.com.

p. 48, **Palantir Technologies is perhaps the most notorious:** Julie Bort, "The Valley's Most Secretive Startup, Palantir, Booked $1.7 Billion in Revenue in 2015 but May Not Be Profitable," *Business Insider*, July 21, 2016, uk.businessinsider.com.

p. 48, **a darling of the US law-enforcement:** Shane Harris, "Palantir Technologies Spots Patterns to Solve Crimes and Track Terrorists," *Wired*, July 31, 2012, wired.co.uk.

p. 48, **Its biggest customers:** Michal Lev-Ram, "Palantir Connects the Dots with Big Data," *Fortune*, March 9, 2016, fortune.com.

p. 48, **it is helping ... home and work addresses:** Spencer Woodman, "Palantir Enables Immigration Agents to Access Information from the CIA," *Intercept*, March 17, 2017, theintercept.com.

p. 48–9, **The most valuable businesses:** Peter Theil, *Zero to One*, Crown Business, 2014, chapter 12.

p. 49, **Now, what they will argue:** Last Week Tonight with John Oliver, "Government Surveillance," April 5, 2015, youtube.com.

p. 49, **It makes it harder:** Stoycheff, Burgess and Martucci, "Online Censorship and Digital Surveillance," 4.

pp. 49–50, **Only when we believe that nobody else is watching:** Greenwald, *No Place to Hide*, "The Harm of Surveillance."

p. 50, **the emphasis on the technical ... technical progress and sophistication:** S. Gürses, A. Kundani and J. van Hoboken, "Crypto and Empire: The Contradictions of Counter-Surveillance Advocacy," *Media, Culture and Society* 38:4, April 2016, 586.

p. 51, **This is a phenomenon:** Stoycheff, Burgess and Martucci, "Online Censorship and Digital Surveillance," 3.

p. 51, **Everything that ... execution of their duty:** Patrick Colquhoun, *A Treatise on the Commerce and Police of the River Thames*, 6th ed., Baldwin and Sons, 1800, 385.

p. 51, **one of the foundational myths:** David Garland, "The Limits of the Sovereign State: Strategies of Crime Control in Contemporary Society," *British Journal of Criminology*, Autumn 1996, 448.

p. 52, **the accepted wisdom:** In his foundational textbook on policing, Robert Reiner describes how "modern societies are characterized by 'police fetishism,' the ideological assumption that the police are a functional prerequisite of social order and that without a police force chaos would ensue." Robert Reiner, *Politics and the Police*, Oxford University Press, 2010, 4.

p. 52, **Academic experiments on this:** Microsoft, for example, announced it would be working with police in developing technology for predictive policing purposes in 2015. Kirk Arthur, "Supporting Law Enforcement Resources with Predictive Policing," Microsoft Blog, January 7, 2015, enterprise.microsoft.com.

p. 52, **a study from UCLA:** Stuart Wolpert, "Predictive Policing Substantially Reduces Crime in Los Angeles during Months-long Test UCLA-led Study Suggests Method Could Succeed in Cities Worldwide," UCLA Newsroom, October 7, 2015, newsroom.ucla.edu.

p. 52, **consultancies selling predictive policing technology:** See, for example, HunchLab, hunchlab.com, and PredPol, predpol.com. See also Maurice Chammah and Mark Hansen, "Policing the Future," Marshall Project Blog, February 3, 2016, themarshallproject.org.

p. 52, **one of the primary data inputs:** See Pedro Burgos, "Highlights from Our Justice Talk on Predictive Policing," Marshall Project Blog, February 25, 2016, themarshallproject.org.

p. 53, **a tendency to target low-income communities:** William Isaac and Andi Dixon, "Why Big-Data Analysis of Police Activity Is Inherently Biased," *Conversation*, May 10, 2017, theconversation.com.

p. 53, **Similar concerns exist with stop-and-frisk policies:** *Floyd et al. v. City of New York, et al.*, Center for Constitutional Rights, Case Summary: ccrjustice.org.

p. 53, **Law enforcement agencies hold records:** *Perpetual Line-Up*, October 16, 2016, perpetuallineup.org.

p. 53, **There are serious, documented issues:** Steve Lohr, "Facial Recognition Is Accurate, if You're a White Guy," *New York Times*, February 9, 2018, nytimes.com.

p. 53, **Both of these:** George Joseph, "The LAPD Has a New Surveillance Formula, Powered by Palantir," *Appeal*, May 8, 2018, theappeal.org.

p. 53, **uncritically ingests yesterday's mistakes:** James Bridle, *New Dark Age: Technology Knowledge and the end of the Future*, Verso, 2018, 144.

p. 54, **academic research suggests that white-collar crime:** Joseph Heath, "The VW Scandal and Corporate Crime," *In Due Course*, October 2, 2015, induecourse.ca.

p. 54, **see this phenomenon in action:** "SKYNET: Courier Detection via Machine Learning," *Intercept*, May 8, 2015, theintercept.com.

pp. 54–5, **documents revealed by Snowden:** Ibid., see also Christian Grothoff and J. M. Porup, "The NSA's SKYNET Program May Be Killing Thousands of Innocent People," *Ars Technica*, February 16, 2016, arstechnica.co.uk.

p. 55, **Zaidan is not a terrorist:** See Ahmad Zaidan, "Al Jazeera's A. Zaidan: I Am a Journalist Not a Terrorist," Al Jazeera, May 15, 2015, aljazeera.com.

p. 55, **The allegations against me:** Ibid.

p. 55, **President Bush discussed targeting Al Jazeera:** Wadah Khanfar, "They Bombed Al Jazeera's Reporters. Now the US Is after Our Integrity," *Guardian*, December 11, 2010, theguardian.com; Matt Wells, "Al Jazeera Accuses US of Bombing Its Kabul Office," *Guardian*, November 17, 2001, theguardian.com.

p. 56, **unreconstructed logical positivism:** Greenfield, *Radical Technologies*, 52.

p. 56, **It also relies:** Bridle, *New Dark Age*, 40.

p. 56, **that sees all social problems:** Vitale, *The End of Policing*, 27, 29.

p. 57, **fundamental rights within the criminal justice system:** See *Civil Rights, Big Data and Our Algorithmic Future*, September 2014, bigdata.fairness.io, chapter 3, "Criminal Justice."

p. 57, **arbitrary interferences with private life:** See Andrew Guthrie Ferguson, "Predictive Policing and Reasonable Suspicion," *Emory Law Journal*, 62:2, 2012–13, law.emory.edu.

p. 57, **discovering their biases and appreciating their nuance:** Greenfield, *Radical Technologies*, 252.

p. 57, **this type of data analysis is used to inform government spending:** Burgos, "Highlights from Our Justice Talk on Predictive Policing." See also Isaac and Dixon, "Why Big-Data Analysis of Police Activity Is Inherently Biased."

p. 58, **Under Ceasefire:** Lois Beckett, "How the Gun Control Debate Ignores Black Lives," ProPublica, November 24, 2015, propublica.org.

p. 58, **Palantir used the city:** Ali Winston, "Palantir Has Secretly Been Using New Orleans to Test Its Predictive Policing Technology," *Verge*, February 27, 2018, theverge.com.

p. 58, **It's supposed to be ran:** Ibid.

p. 59, **We are developing ... welfare-ism and social citizenship:** Garland, "The Limits of the Sovereign State," 462.

p. 59, **infinite difficulties and discouragements ... the extensive and enormous evils:** Colquhoun, *A Treatise on the Commerce and Police of the River Thames*, 245.

p. 60, **Microsoft had become aware:** Brad Smith, "The Need for Urgent Collective Action to Keep People Safe Online: Lessons from Last Week's Cyberattack," Microsoft Blog, May 14, 2017, blogs.microsoft.com.

p. 61, **When it was discovered:** Philip N. Howard, *Pax Technica*, Yale University Press, 2015, "Building a Democracy of Our Own Devices."

p. 61, **It had been identified by the NSA:** Ibid.; Kim Zetter, "Report: NSA Exploited Heartbleed to Siphon Passwords for Two Years," *Wired*, April 11, 2014, wired.com.

p. 61, **This has never been a conversation:** Transcript: Edward Snowden speaks at the first annual K(NO)W Identity Conference, May 15, 2017, oneworldidentity.com.

p. 61, **the early years of a cyberwar arms race:** Bruce Schneier, "Cyberwar Treaties," *Crypto-Gram*, June 15, 2012, schneier.com.

p. 63, **The proverbial small spark:** *Evening News and Post*, August 26, 1889, portcities.org.uk.

p. 65, **He suffered permanent:** *Grimshaw v. Ford Motor Co.* (1981) Cal. App. 3d, vol. 119, law.justia.com.

p. 65, **The body of the car reportedly:** Richard T. De George, "Ethical Responsibilities of Engineers in Large Organizations: The Pinto Case," *Business and Professional Ethics Journal* 1:1 (Fall 1981), 1. See also John R. Danley, "Polishing up the Pinto: Legal Liability, Moral Blame, and Risk," *Business Ethics Quarterly* 15:2 (April 2005), 208ff.

p. 66, **Burn injuries and burn deaths:** Mike Dowie, "Pinto Madness," *Mother Jones*, October 1977, motherjones.com.

p. 66, **Ford knows the Pinto is a firetrap:** Dowie, "Pinto Madness."

p. 66, **It is hard to imagine:** Henry Petroski, "Engineering: Backseat Designers," *American Scientist* 100:3 (May–June 2012), 194–5.

p. 66, **Self-styled experts ... wishful thinking:** Ralph Nader, *Unsafe at Any Speed*, Grossman, 1972, 3–4.

p. 66, **W. B. Murphy, president:** Ibid., xxxiii.

p. 67, **key positions in regulatory agencies:** Ibid., xxvii.

p. 67, **The Federal funding:** Ibid., xxxi–xxxii.

p. 67, **one of the biggest industries:** Ibid., 324–5.

p. 67, **a neglected epidemic:** Petroski, "Engineering: Backseat Designers," 194.

p. 67, **estimates vary from hundreds to thousands:** Dowie, "Pinto Madness"; see also Danley, "Polishing Up the Pinto," 206.

p. 67, **GM's Corvair:** Nader, *Unsafe at Any Speed*, 9.

p. 67, **began with the conception:** Ibid., 19.

p. 69**, a non-reformist reform:** Nick Srnicek and Alex Williams, *Inventing the Future: Postcapitalism and a World without Work*, Verso, 2015, "Building Power."

p. 69, **Latanya Sweeney—Arrested?:** Latanya Sweeney, "Discrimination in Online Ad Delivery," *Communications of the ACM* 56:5, dataprivacylab.org.

p. 70, **Ads suggesting arrest:** Ibid., 4.

p. 70**, a greater percentage of ads:** Ibid., 22.

pp. 70–1, **Google understands that an advertiser:** Ibid., 34.

p. 71, **commonly held form of implicit bias:** Chris Mooney, "The Science of Why Cops Shoot Young Black Men," *Mother Jones*, December 1, 2014, motherjones.com.

p. 73, **blithely generate:** Cathy O'Neil, *Weapons of Math Destruction*, Crown, 2016, "Sweating Bullets: On the Job."

p. 73 **Managers assume that … accurate or not:** Ibid.

p. 73, **biased against people:** Ibid., "Ineligible to Serve."

p. 73, **a predicted score as a proxy:** Indiana University Bloomington, "New Research Uncovers Hidden Bias in College Admissions Tests," January 25, 2016, news.indiana.edu releases.

p. 74, **Algorithms are also used:** Nicholas Diakopoulos, "We Need to Know the Algorithms the Government Uses to Make Important Decisions about Us," *Conversation*, May 24, 2016, theconversation.com.

p. 74**, photo of some black people as gorillas:** Jessica Guynn, "Google Photos Labeled Black People 'Gorillas'," *USA Today*, July 1, 2015, usatoday.com.

p. 74 **The privileged:** O'Neil, *Weapons of Math Destruction*, Introduction.

p. 74, **a disproportionate amount of junk food:** Josh Harkinson, "Walmart Ads Target "Low Income" Consumers with Junk Food," *Mother Jones*, November 13, 2013, motherjones.com.

p. 74**, software may pre-emptively exclude:** Kaveh Waddell, "How Big Data Harms Poor Communities," *Atlantic*, April 8, 2016, theatlantic.com.

p. 74, **our current conversations:** Olivia Solon, "Elon Musk: We Must Colonise Mars to Preserve Our Species in a Third World War," *Guardian*, March 11, 2018, theguardian.com.

p. 75**, perhaps for them:** Kate Crawford, "Artificial Intelligence's White Guy Problem," *New York Times*, June 25, 2016, nytimes.com.

p. 75, **contemporary technologies never work:** Greenfield, *Radical Technologies*, 299.

p. 75, **Five groups have a higher than average:** Pew Trusts, *Payday Lending in America*, 2012, pewtrusts.org, 4.

p. 75, **80 percent of payday loans:** Consumer Finance Protection Bureau, *CFPB Data Point: Pay Day Lending*, March 2014, consumerfinance.gov.

p. 75, **Payday lenders find customers:** Upturn, *Led Astray: Online Lead Generation and Payday Loans*, October 2015, upturn.com.

p. 75, **In response to lobbying efforts:** David Graff, "An Update to Our Adwords Policy on Lending Products," *Google Blog*, May 11, 2016, blog.google.

p. 76, **luring students to enroll:** US Senate Committee Report, *For Profit Higher Education: The Failure to Safeguard the Federal Investment and Ensure Student Success*, 2012, gpo.gov, 758.

p. 76, **more than half of students:** Ibid., 1–4.

p. 76, **Welfare Mom w/Kids:** Ibid., 766.

p. 76, **A potential student's first click:** O'Neil, *Weapons of Math Destruction*, "Propaganda Machine: Online Advertising."

p. 76, **The for-profit colleges:** Ibid.

p. 77, **privileges the powers of data scientists:** Turow, *The Daily You*, Introduction.

p. 77, **a modern equivalent of redlining:** Julianne Tveten, "Digital Redlining: How Internet Service Providers Promote Poverty," *Truthout*, December 14, 2016, truth-out.org.

p. 77, **ethnic affinities:** Julia Angwin and Terry Parris Jr., "Facebook Lets Advertisers Exclude Users by Race," *ProPublica*, October 28, 2016, propublica.org.

p. 77, **They have also admitted:** Julia Angwin, Noam Scheiber, and Ariana Tobin, "Dozens of Companies Are Using Facebook to Exclude Older Workers from Job Ads," ProPublica, December 20, 2017, propublica.org.

p. 77, **agreed to voluntarily recall:** Danley, "Polishing Up the Pinto," 209.

p. 77, **the conduct of Ford's management:** *Grimshaw v. Ford Motor Co.* (1981), 119 Cal. App. 3d 819.

p. 78, **Code itself functions:** Lawrence Lessig, *Code Version 2.0*, Basic Books, 2006, 5. Lessig describes this as code being a "salient regulator."

p. 78, **Code is never found:** Ibid., 6.

p. 78, **structural inequalities of the past:** Ian Tucker, "A White Mask Worked Better: Why Algorithms Are Not Colour Blind," *Observer*, May 28, 2017, theguardian.com.

p. 79, **A democratic government:** Nader, *Unsafe at Any Speed*, 332–3.

p. 79, **Importantly, workers who build:** Lizzie O'Shea, "Tech Has No Moral Code. It Is Everyone's Job Now to Fight for One," *Guardian*, April 26, 2018, theguardian.com.

p. 79, **Technological professionals are ... like that again:** Cherri M. Pancake, "Programmers Need Ethics When Designing the Technologies that Influence People's Lives," *Conversation*, August 8, 2018, theconversation.com.

pp. 79–80, **As the people who build:** Letter to Satya Nadella, int.nyt.com; Sheera Frenkel, "Microsoft Employees Protest Work with ICE, as Tech

Industry Mobilizes over Immigration," *New York Times*, June 19, 2018, nytimes.com.

p. 80, **Similar insurgencies:** Mark Bergen, "Google Engineers Refused to Build Security Tool to Win Military Contracts," *Bloomberg*, June 21, 2018, bloomberg.com.

p. 81, **Amazing progress is being made:** Jordan Kisner, "How a New Technology Is Changing the Lives of People Who Cannot Speak," *Guardian*, January 23, 2018, theguardian.com.

p. 81, **Advances in technology:** Elise Thomas, "Why the Internet of Things Matters in the Fight against Domestic Violence," November 17, 2015, medium.com.

p. 82, **More than a third of women:** National Center for Injury Prevention and Control of the Centers for Disease Control and Prevention, *The National Intimate Partner and Sexual Violence Survey: 2010 Summary Report*, 2010, cdc.gov.

p. 82, **97 percent report:** National Network to End Domestic Violence, "A Glimpse from the Field: How Abusers Are Misusing Technology," 2014, static1.squarespace.com.

p. 82, **Nearly half of the shelters:** Aarti Shahani, "Smartphones Are Used to Stalk, Control Domestic Abuse Victim," *All Tech Considered*, September 15, 2014, npr.org.

p. 83, **In the future:** Spencer Ackerman and Sam Thielman, "US Intelligence Chief: We Might Use the Internet of Things to Spy on You," *Guardian*, February 9, 2016, theguardian.com.

p. 83, **In case you are wondering:** Evgeny Morozov, twitter.com.

p. 83, **About $166 was spent:** Nader, *Unsafe at Any Speed*, 337–8.

p. 84, **not employ a single black woman:** Reveal (Center for Investigative Reporting), "Here's the Clearest Picture of Silicon Valley's Diversity Yet: It's Bad. But Some Companies Are Doing Less Bad," June 25, 2018, revealnews.org.

p. 84, **But white people:** US Equal Employment Opportunity Commission, *Diversity in High Tech*, May 2016, eeoc.gov.

p. 84, **These propositions become:** Greenfield, *Radical Technologies*, 41.

p. 84, **less about real innovation:** Olivia Solon, "CES 2018: Less 'Whoa,' More 'No!'—Tech Fails to Learn From Its Mistakes at Annual Pageant," *Guardian*, January 10, 2018, theguardian.com.

p. 84, **[San Francisco] tech culture:** Overheard by Twitter user Aziz Shamim, twitter.com.

p. 85, **The Google walkout:** Claire Stapleton, Tanuja Gupta, Meredith Whittaker, Celie O'Neil-Hart, Stephanie Parker, Erica Anderson, and Amr Gaber, "We're the Organizers of the Google Walkout. Here Are Our Demands," *Cut*, November 1, 2018, thecut.com; Michael Walker, "The Google Walkout Is a Watershed Moment in 21st Century Labour Activism," *Conversation*, November 8, 2018, theconversation.com.

p. 85, **The lack of diversity:** See Catherine Ashcroft, "10 Actionable Ways to Actually Increase Diversity in Tech," *Fast Company*, January 26, 2015, fastcompany.com.

p. 85 **educational programs on ethics:** Nathasha Singer, "Tech's Ethical 'Dark Side': Harvard, Stanford and Others Want to Address It," *New York Times*, February 12, 2018, nytimes.com; Simone Stolzoff, "Are Universities Training Socially Minded Programmers?," *Atlantic*, July 24, 2018, theatlantic.com.

p. 86, **be aware that if your spoken words:** Dan Graziano, "Disable This Feature to Stop Your Samsung Smart TV from Listening to You," *Cnet*, February 10, 2015, cnet.com.

p. 86, **It's just a matter of time:** Whitney Meers, "Hello Barbie, Goodbye Privacy? Hacker Raises Security Concerns," *Huffington Post*, November 30, 2015, huffingtonpost.com.

p. 87, **In his analysis:** Nader, *Unsafe at Any Speed*, 333.

p. 88, **This can happen with objects:** Matt Shaer, "The False Promise of DNA Testing," *Atlantic*, June 2016, theatlantic.com.

p. 88, **results that were later discredited:** Ibid.

p. 88, **defense lawyers objected:** Andrew Keshner, "Judge Rejects Medical Examiner's DNA Technique," *New York Law Journal*, July 15, 2015, legal-aid.org.

p. 88, **legal aid lawyers managed:** Renate Lunn, "The Public Defender's Fight for Justice," *New York Law Journal*, March 21, 2017, legal-aid.org.

p. 88, **the judge decided:** Keshner, "Judge Rejects Medical Examiner's DNA Technique."

p. 89, **LRMix Studio offers an alternative:** See lrmixstudio.org.

p. 89, **Similar open source tools:** See Corina C. G. Benscho, Jeroen de Jong, Linda van de Merwe, Vanessa Vanvooren, Morgane Kempenaers, Kees van der Beek, Filippo Barni, Eusebio López Reyes, Léa Moulin, Laurent Pene, Peter Gill, Titia Sijen, Hinda Haned, "SmartRank: An Open Source Likelihood Ratio Software for Searching National Databases with Complex DNA Profiles," International Symposium on Human Identification News, July 29, 2017, ishinews.com.

p. 89, **There are growing calls:** *AI Now 2017 Report*, assets.contentful.com. See also Tom Simonite, "AI Experts Want to End 'Black Box' Algorithms in Government," *Wired*, October 18, 2017, wired.com.

p. 90, **since then, the market:** Cade Metz, "The Battle for Top AI Talent Only Gets Tougher from Here," *Wired*, March 23, 2017, wired.com.

p. 90, **important recent initiatives:** See partnershiponai.org; and see Tim Simonte, "As Artificial Intelligence Advances, Here Are Five Tough Projects for 2018," *Wired*, December 21, 2017, wired.com; James Vincent, "Elon Musk, DeepMind Founders, and Others Sign Pledge to Not Develop Lethal AI Weapon Systems," *The Verge*, July 18, 2018, theverge.com.

p. 91, **concentrations of power:** Katharine Dempsey, "Democracy Needs a Reboot for the Age of Artificial Intelligence," *Nation*, November 8, 2017, thenation.com.

p. 91, **the NHTS increased the speed:** Matthew T. Lee, "The Ford Pinto Case and the Development of Auto Safety Regulations, 1893–1978," *Business and Economic History* 27:2 (Winter 1998), 391.

p. 92, **in one significant way:** David E. Broockman, Gregory Ferenstein, Neil Malhotra, "Wealthy Elites' Policy Preferences and Economic Inequality: The Case of Technology Entrepreneurs," Working Paper, September 5, 2017, 4–6, gsb.stanford.edu.

p. 93, **The regulation of the automobile:** Nader, *Unsafe at Any Speed*, 343.

p. 95, **I called in Dr. Pillsbury:** Edward Bellamy, *Looking Backward*, Project Gutenberg, gutenberg.com, 14.

pp. 95–6, **Every quarter contained:** Ibid., 14–19.

p. 96, **the most widely read:** Bertell Ollman, "The Utopian Vision of the Future (Then and Now): A Marxist Critique," *Monthly Review* 57:3 (July–August 2005), monthlyreview.org.

p. 97, **how men with children:** Bellamy, *Looking Backward*, 53.

p. 97, **All that society had to do:** Ibid., 23.

p. 97, **federal system of autonomous nations:** Ibid., 56, 28.

p. 98, **poverty was impossible ... a map of the world:** Oscar Wilde, *The Soul of Man under Socialism*, Arthur L. Humphreys, 1900, 40.

p. 98, **technological progress as the means:** Howard P. Segal, *Technological Utopianism in American Culture*, Syracuse University Press, 2005, 10.

p. 98, **technology serves as a remedy:** Joseph Corn, *Imagining Tomorrow: History, Technology, and the American Future*, MIT Press, 1986, 219–23; and Howard P. Segal, "The Technological Utopians," in Corn, *Imagining Tomorrow*, 123, 130.

p. 98, **the specific kind of thinking:** Segal, *Technological Utopianism in American Culture*. In his seminal work, Segal examines a range of texts by technological utopians from this period.

p. 98, **the most important contemporary trend:** Segal, *Technological Utopianism in American Culture*, 21.

p. 99, **even the weather:** John Macnie, *The Diothas, or, A Far Look Ahead*, Putnam, 1883, 123.

p. 99, **large-scale machinery:** Henry Olerich, *A Cityless and Countryless World: An Outline of Practical Co-operative Individualism*, Gilmore & Olerich, 1893, 122; Herman Hine Brinsmade, *Utopia Achieved: A Novel of the Future*, Broadway Publishing, 1912, 39–40.

p. 99, **personalized, compact electric motors:** Macnie, *The Diothas*, 15, 21.

p. 99, **glass with the strength:** Olerich, *A Cityless and Countryless World*, 53.

p. 99, **Communal kitchens:** Macnie, *The Diothas*, 43–51; Olerich, *A Cityless and Countryless World*, 86–9.

p. 99, **a movement not of revolt:** Segal, "The Technological Utopians," 120.

p. 100, **greater intelligence among the masses:** Olerich, *A Cityless and Countryless World*, 97, 110.
p. 100, **recognition of and cooperation with industrial evolution:** Bellamy, *Looking Backward*, 23; see also Segal, *Technological Utopianism in American Culture*, 22.
p. 100, **the unstoppable expansion of corporations:** King Camp Gillette, *World Corporation*, New England News Company, 1910, 9. Gillette proposes a plan for transformation that is similar to what is described in Bellamy's fictional account, in which a universal system of industrial organization comes about through gradual absorption of small companies by large ones, 131. Jack London, *The Human Drift*, Macmillan, 1894.
p. 100, **the transformative ability to manufacture power:** See George S. Morison, *The New Epoch as Developed by the Manufacture of Power*, Arno Press, 1903.
p. 100, **[the engineer] is the priest:** Ibid., 75–6.
p. 101, **cogs in the machine:** Gillette, *World Corporation*, 132.
p. 101, **If any live in idleness:** Macnie, *The Diothas*, 12.
p. 101, **neglect of work:** Bellamy, *Looking Backward*, 50.
p. 101, **We teach that labour:** Olerich, *A Cityless and Countryless World*, 74.
p. 101, **immense harm:** Bertrand Russell, *In Praise of Idleness*, Allen & Unwin, 1935.
p. 101, **licentiousness considered "unchaste":** Olerich, *A Cityless and Countryless World*, 273–4.
p. 101, **sexual mores remained rigid:** Segal, *Technological Utopianism in American Culture*, 31.
p. 102, **It is their privilege … mainly housework:** Macnie, *The Diothas*, 18, 45–8.
p. 102, **There will be no voting:** Gillette, *World Corporation*, 54.
p. 102, **Therefore, popular voting:** Harold Loeb in *Life in a Technocracy*, Syracuse University Press, 1933, 75.
pp. 102–3, **has power to join in harmony:** Gillette, *World Corporation*, 9, 132.
p. 103, **The most serious obstruction:** Ibid., 176.
p. 103, **The nation … :** Bellamy, *Looking Backward*, 25.
p. 103, ***free* competition, of *healthy* supply:** Olerich, *A Cityless and Countryless World*, 163.
p. 104, **As techno-fixes to political problems:** Segal, *Technological Utopianism in American Culture*, 165.
p. 104, **run headlong:** Alec Berg speaking on *Fresh Air*, NPR, June 9, 2016.
p. 105, **I don't know about you people:** *Silicon Valley*, season 2, episode 1.
p. 105, **infinite data … solve big problems:** Eric Schmidt and Jonathan Rosenberg, *How Google Works*, Grand Central Publishing, 2017, "Big Problems Are Information Problems."
p. 105, **Bernard Harcourt has also written:** Harcourt, *Exposed*, 127–8.
p. 105, **the most complete picture:** Ibid., 128.

p. 105, **their influential essay:** Richard Barbrook and Andy Cameron, "The Californian Ideology," *Mute* 1:3 (September 1, 1995), metamute.org.

p. 106, **promiscuously [combine] the free-wheeling spirit:** This is from a later incarnation of Richard Barbrook and Andy Cameron, "The Californian Ideology," *Science as Culture* 6:t1 (1996), 44.

p. 106, **the boyish air of a nascent superhero:** Meaghan Daum, "Elon Musk Wants to Change How (and Where) Humans Live," *Vogue*, September 21, 2015, vogue.com.

p. 106, **He's extremely smart:** Lev Grossman, "Inside Facebook's Plan to Wire the World," *Time*, December 15, 2014, time.com.

p. 107, **This is perhaps unsurprising:** Jane Bird, "How the Tech Industry Is Attracting More Women," *Financial Times*, March 9, 2018, ft.com.

p. 107, **repeated removal of images of breastfeeding:** Nicola Bartlett, "Facebook Bans ANOTHER Breastfeeding Photo—after Complaints from Men about One Nipple," *Mirror*, 13 February 2015, mirror.co.uk.

p. 107, **a fat woman in a bikini:** Sam Levin, "Too Fat for Facebook: Photo Banned for Depicting Body in 'Undesirable Manner'," *Guardian*, May 23, 2016, theguardian.com.

p. 107, **censorship of photos of Australian Aboriginal women:** Kevin Rennie, "'Nude' Photos of Australian Aboriginal Women Trigger Facebook Account Suspensions," *Global Voices Vox*, March 16, 2016, advox.globalvoices.org.

p. 107, **iconic Pulitzer Prize–winning photograph:** Julie Carrie Wong, "Mark Zuckerberg Accused of Abusing Power after Facebook Deletes 'Napalm Girl' Post," *Guardian*, September 9, 2016, theguardian.com.

p. 107, **Regulations get created in anticipation of problems:** Schmidt and Rosenberg, *How Google Works*, "Big Problems Are Information Problems," "The Role of Government."

p. 108, **the disease of being policy wonks:** Michiko Kakutani, "Glimpses of Obama among Friends," *New York Times*, September 18, 2011, nytimes.com.

p. 108, **to protect this precious democracy:** Andrew Sullivan, "Democracies End When They Are Too Democratic," *New York Magazine*, May 2, 2016, nymag.com.

p. 109, **build an opt-in society:** Balaji Srinivasan, "Silicon Valley's Ultimate Exit," genius.com.

p. 109, **the settlement of the New World:** Christian Davenport, "Elon Musk Provides New Details on His 'Mind Blowing' Mission to Mars," *Washington Post*, June 10, 2016, washingtonpost.com.

p. 109, **no longer believe[s] that freedom:** Peter Thiel, "The Education of a Libertarian," April 13, 2009, *Cato Unbound*, cato-unbound.org.

p. 110, **I'm not exactly sure:** Kyle DeNuccio, "Silicon Valley Is Letting Go of Its Techie Island Fantasies," *Wired*, May 16, 2015, wired.com.

p. 112, **an autonomously organized society:** See Kristin Ross, *Communal Luxury*, Verso, 2015, 1.

p. 112, **A full 84 percent:** William A. Pelz, *A People's History of Modern Europe*, Pluto, 2016, 79.

p. 112, **You are masters of your destiny:** "Your Commune Has Been Constituted," March 29, 1871, marxists.org.

p. 112, **The recognition and consolidation:** Rev. William Gibson, *Paris During the Commune*, Haskell House, 1974, 212–13.

p. 113, **The Commune ... night work for bakers:** See Juliet Jacques, "Returning to the Commune of Paris," *New Statesman*, November 13, 2012, newstatesman.com; Paul D'Amato, "The Paris Commune," *Socialist Worker*, December 13, 2002; Flick Ruby, "Louise Michel," spunk.org/library.

p. 113, **Workers began to organize:** John Merriman, *Massacre—The Life and Death of Paris*, Basic Books, 2014, 63; Pelz, *A People's History of Modern Europe*, 80.

p. 113, **fair remuneration for their work:** Gibson, *Paris during the Commune*, 136.

p. 113, **They formed a *Union des Femmes*:** Merriman, *Massacre*, 66.

p. 113, **It was the Commune's largest:** Ross, *Communal Luxury*, 27.

p. 113, **Women achieved positions of power:** Pelz, *A People's History of Modern Europe*, 81.

p. 113, **Marriage laws were liberalized:** Merriman, *Massacre*, 61, 66.

p. 113, **A Federation of Artists:** Ibid., 54.

p. 113, **Paris is a true paradise:** Ibid., 53.

p. 114, **did not expect miracles:** Karl Marx, "The Paris Commune" [1871], marxists.org.

p. 114, **any conditions which arise:** Ollman, "The Utopian Vision of the Future."

p. 114, **The Communards had not decreed:** Ross, *Communal Luxury*, 79.

p. 115, **the Commune was crushed:** Violence, it must be said, was ever-present during the life of the Commune, but any comparison between the Communards and the French generals makes clear that it was not evenly distributed. Even in the beginning, when the Thiers government sought to disarm the Parisians by confiscating their cannons, several French generals were shot as well as National Guardsmen in the fracas that followed. Carolyn J. Eichner, *Surmounting the Barricades: Women in the Paris Commune*, Indiana University Press, 2004, 25; Merriman, *Massacre*, 26–9.

p. 115, **The "Bloody Week":** Eichner, *Surmounting the Barricades*, 35.

p. 115, **The virulence of Versailles's:** Ibid.

p. 115, **modestly titled profile:** Chris Anderson, "The Man Who Makes the Future: Wired Icon Marc Andreesson," *Wired*, April 24, 2012, wired.com.

p. 115, **an evangelist for the church:** Tad Friend, "Tomorrow's Advance Man," *New Yorker*, May 18, 2015, newyorker.com.

p. 116, **I am not talking about Marxism:** Marc Andreesson, "This Is Probably a Good Time to Say That I Don't Believe Robots Will Eat All the Jobs," June 13, 2014, blog.pmarca.com.

p. 116, **Andreesson sees many of our collective problems:** Friend,

"Tomorrow's Advance Man." "When I brought up the raft of data suggesting that intra-country inequality is in fact increasing, even as it decreases when averaged across the globe—America's wealth gap is the widest it's been since the government began measuring it—Andreessen rerouted the conversation, saying that such gaps were 'a skills problem,' and that as robots ate the old, boring jobs humanity should simply retool."

p. 116, **Firms trumpet their boldness:** Ibid.

p. 116, **Utopian thinking presents us:** Ollman, "The Utopian Vision of the Future."

p. 117, **This analysis underpinned:** William Morris, "Bellamy's *Looking Backward*," *Commonweal*, June 21, 1889.

p. 117, **foremost British supporters:** Ross, *Communal Luxury*, 61.

p. 118, **to lead the Web to its full potential:** See w3.org/Consortium.

p. 119, **The W3C is by no means perfect:** Prabir Purkayastha and Rishab Bailey, "US Control of the Internet," *Monthly Review* 66:3 (July–August, 2014) monthlyreview.org.

p. 119, **Were all of these tools:** Matt Bruenig, "The Amusing Case of Tech Libertarians," June 6, 2015, mattbruenig.com.

p. 119, **gigantic in scale:** Gary Wolf, "Why Craigslist Is Such a Mess," *Wired*, August 24, 2009, wired.com.

p. 119, **tiny competitors in the search engine market:** Charles Arthur, "NSA Scandal Delivers Record Numbers of Internet Users to DuckDuckGo," *Guardian*, July 10, 2013, guardian.com.

p. 120, **are developed, maintained, and upgraded:** Rebecca MacKinnon, *Consent of the Networked: The Worldwide Struggle for Internet Freedom*, Basic Books, 2012, 231.

p. 120, **The Tech Workers Coalition:** Ben Tarnoff, "Trump's Tech Opposition," *Jacobin*, May 2, 2017, jacobinmag.com. They are not the only such organization; see Chi Onwurah, "All Workers Need Unions—Including Those in Silicon Valley," *Guardian*, January 31, 2018, guardian.com.

p. 120, **Similar groups are springing up:** Vinita Govindarajan, "From War Protestors to Labour Activism: India's First IT Workers Union Is Being Formed in Tamil Nadu," May 22, 2017, *Scroll.In*, scroll.in.

p. 121, **Above all, I hope she is not *poetical*:** Sydney Padua, *The Thrilling Adventures of Lovelace and Babbage*, Pantheon Books, 2015, 34.

pp. 121–2, **We desire certainty:** Ibid., 34.

p. 122, **The reason is obvious:** Ibid., 267, quoting a letter from Augustus De Morgan to Lady Byron on Lady Lovelace's mathematical education.

pp. 122–3, **turning the transcription:** Ibid., 25.

p. 123, **the direct praise bestowed on her work:** Luigia Carlucci Aiello, "The Multifaceted Impact of Ada Lovelace in the Digital Age," *Artificial Intelligence*, 235 (2016), 60.

p. 123, **by manipulating symbols:** Padua, *The Thrilling Adventures of Lovelace and Babbage*, 26.

p. 123, **The stubborn, rigid Babbage:** Ibid., 25.

p. 124, **Enchantress who has thrown her magical spell:** James Essinger, *Ada's Algorithm: How Lord Byron's Daughter Ada Lovelace Launched the Digital Age*, Melville House, 2014, xiv.

p. 124, **Invert the order!:** Padua, *The Thrilling Adventures of Lovelace and Babbage*, 27.

p. 124, **He rejected her overtures:** Essinger, *Ada's Algorithm*, 184–92.

p. 124, **technology is a very human activity:** Melvin Kranzberg, "Presidential Address, Technology and History: 'Kranzberg's Laws,'" *Technology and Culture* 27:3 (1986), 557.

p. 125, **As machines were built to do work:** This idea is derived from a quote that appeared on the wall of an exhibition on robots at the Museum of Science in London (2017): "When we built machines to do human work, human workers felt like robots."

p. 126, **Luddites opposed the use of machines:** Kevin Binfield, ed., *Writings of the Luddites*, Johns Hopkins University Press, 2004, 3.

p. 126, **They objected to machinery:** Richard Coniff, "What the Luddites Really Fought Against," *Smithsonian Magazine*, March 2011, smithsonianmag.com.

p. 126, **have arisen from circumstances ... and their community:** Maiden Speech of Lord Byron, 1812, as featured in Robert Charles Dallas, *Recollection of the Life of Lord Byron, from the year 1808 to the end of 1814*, lordbyron.org, 206.

p. 126, **The Remedy for you:** Binfield, ed., *Writings of the Luddites*, 1.

p. 127, **Man produces himself:** Robert Tucker, ed., *The Marx–Engels Reader*, Norton, 1978, 76.

p. 127 **resolved personal worth into exchange value:** Marx and Engels, *Manifesto of the Communist Party*, chapter 1, marxists.org.

p. 128, **For those observing the development:** Amy E. Wendling, *Karl Marx on Technology and Alienation*, Palgrave Macmillan, 2009, 2.

p. 128, **In tearing away from man:** Karl Marx, *Economic and Philosophical Manuscripts of 1844*, "Estranged Labour," marxists.org.

p. 128, **Amy Wendling notes:** Wendling, *Karl Marx on Technology and Alienation*, 2.

p. 128, **a step, if treacherous, towards liberation:** Ibid., 9.

p. 128, **almost total intellectual isolation:** Raúl Rojas, "Encyclopaedia of Konrad Zuse," *Konrad Zuse Internet Archive*, zuse.zib.de.

p. 129, **IBM was one such example:** James Bessen, "What Good Is Free Software?" in Robert Hahn, ed., *Is Open Source the Future of Software?*, AEI-Brookings Joint Center for Regulatory Studies, 2002, 14.

p. 130, **mainframe software was cooperatively developed:** Eben Moglen, *Anarchism Triumphant*, Columbia University Press, 1999, emoglen.law.columbia.edu.

p. 130, **One computer company ended up:** Stephen Levy, *Hackers: Heroes of* *the Computer Revolution*, Dell Publications, 1984, 53.

p. 131, **Levy describes hacking:** Ibid., 8.

p. 131, **kind of restless curiosity:** Ibid., 8–9.

p. 131, **To qualify as hacking:** Ibid., 10.

p. 131, **He began hanging around:** Ibid., 98–105.

pp. 131–2, **There were women programmers:** Ibid., 72.

p. 132, **Hopper was instrumental:** Claire L. Evans, *Broad Band: The Untold Story of the Women Who Made the Internet*, Penguin Random House, 2018, "Tower of Babel."

p. 132, **As the industry began to professionalize:** Holly Brockwell, "Sorry, Google Memo Man: Women Were in Tech Long Before You," *Guardian*, August 9, 2017, theguardian.com.

p. 133, **these prejudices have proven:** Owen Jones, "Google's Sexist Memo Has Provided the Alt-Right with a New Martyr," *Guardian*, August 8, 2017, guardian.com.

p. 133, **This memo argued:** James Damore, "Google's Ideological Echo Chamber," July 2017, assets.documentcloud.org.

p. 133, **You can create art:** Levy, *Hackers*, 30.

p. 133, **When I speak about computer programming:** Donald Knuth at ACM Turing Award Lecture, *Communications of the ACM* 17:12 (December 1974), 668, 670.

p. 134–5, **IBM "unbundled" its products:** Bessen, "What Good Is Free Software?" 14.

p. 135, **As the majority of hobbyists:** Bill Gates, "Letter to the Hobbyists," *Homebrew Computer Club Newsletter* 2:1 (January 31, 1976).

p. 135, **Some directly opposed it:** See, for example, Jim Warren's letter in response: "There is a viable alternative to the problems raised by Bill Gates in his irate letter to computer hobbyists concerning 'ripping off' software. When software is free, or so inexpensive that it's easier to pay for it than to duplicate it, then it won't be 'stolen'." Jim Warren, "Correspondence," *SIGPLAN Notices* (ACM) 11:7 (July 1976).

p. 136, **Fellow developers sold programs:** Glyn Moody, *Rebel Code*, Penguin, 2002, 18; Johan Söderberg, *Hacking Capitalism: The Free and Open Source Software Movement*, Routledge, 2008, 20.

p. 136, **He called this system GNU:** Moody, *Rebel Code*, 18.

p. 136, **Software sellers want … I refuse to break solidarity:** Richard Stallman, *The GNU Manifesto*, gnu.org.

p. 137, **The Luddites struggled against:** John Zerzan and Paula Zerzan, "Who Killed Ned Ludd? A History of Machine Breaking at the Dawn of Capitalism," *Fifth Estate*, no. 271 (April 1976), fifthestate.org.

p. 137, **Stallman worked with lawyers:** Moody, *Rebel Code*, 26.

p. 137, **to guarantee your freedom:** Preamble to the GNU Public License v3, gnu.org.

p. 137, **The only requirement:** GNU Public License v3, gnu.org.

p. 138, **In 1979, one of the main kernels:** Söderberg, *Hacking Capitalism*, 19, 23; Moody, *Rebel Code*, 33.
p. 139, **In part because of the ambitious size:** Moody, *Rebel Code*, 49.
p. 139, **Making Linux freely available:** Interview with Linus Torvalds, *First Monday Club*, March 2, 1998, firstmonday.org.
p. 139, **Such an environment is reminiscent:** Söderberg, *Hacking Capitalism*, 9.
p. 139, **The result was that Linux:** Ibid., 24.
p. 139, **Everybody puts in effort:** Interview with Torvalds, *First Monday Club*.
p. 140, **The resulting body of software:** See Bessen, "What Good Is Free Software?" in Hahn, ed., *Is Open Source the Future of Software?* 17. This is something explicitly acknowledged by Microsoft, in a series of leaked internal memos in the late 1990s, called the Halloween memos.
p. 141, **But once on the road:** Guilbert Gates, Jack Ewing, Karl Russell and Derek Watkins, "Explaining Volkswagen's Emissions Scandal," *New York Times*, April 28, 2016, nytimes.com.
p. 141, **contain nearly twice the amount:** David Gelles, Hiroko Tabuchi and Matthew Dolan, "Complex Car Software Becomes the Weak Spot under the Hood," *New York Times*, September 26, 2015, nytimes.com.
p. 141, **For the last few years:** Nicole Perlroth, "Security Researchers Find a Way to Hack Cars," *New York Times*, July 21, 2015, bits.blogs.nytimes.com.
p. 141, **plugging into a diagnostic port:** Ibid.
p. 141, **I was driving 70 mph:** Andy Greenberg, "Hackers Remotely Kill a Jeep on the Highway—With Me in It," *Wired*, July 21, 2015, wired.com.
p. 142, **It is deeply ironic:** Alex Davies, "The EPA Opposes Rules That Could've Exposed VW's Cheating," *Wired*, September 18, 2015, wired.com.
p. 143, **scandalous:** Geoff Colvin, "Volkswagen's Scandal Management: Needs Improvement," *Fortune*, September 24, 2015, fortune.com.
p. 143, **a degree of complacency and pomposity:** Graham Ruddick, "Volkswagen: A History of Boardroom Clashes and Controversy," *Guardian*, September 23, 2015, theguardian.com.
p. 143, **accident waiting to happen:** James B. Stewart, "Problems at Volkswagen Start in the Boardroom," *New York Times*, September 24, 2015, nytimes.com.
p. 143, **founded by the Nazis:** Marc Meillassoux, "Mood in Wolfsburg, City of Volkswagen's Headquarters, Gloomier than Ever," ABC News, September 25, 2015, abcnews.go.com.
p. 143, **Mundane now appears to be:** Damian Carrington, "Four More Carmakers Join Diesel Emissions Row," *Guardian*, October 9, 2015, theguardian.com.
p. 143, **It becomes a dangerous building material:** Eben Moglen, "When Software Is in Everything: Future Liability Nightmares Free Software

Helps Avoid," speech given at the Scottish Society for Computers and Law annual meeting, June 30, 2010, softwarefreedom.org.

pp. 143–4, **Free software:** Eric Steven Raymond in *The Cathedral and the Bazaar*, 1997, gets at this when describing the contrast between the "quiet, reverent cathedral-building" of proprietary software with the "great babbling bazaar of differing agendas and approaches" that makes up the Linux community. Unfortunately, Raymond has since gone on to express some appalling views on a range of topics, which undermines his credibility but does not diminish the quality of this text as a historical artifact. See Sarah Jeong, "Meet the Campaign Connecting Affluent Techies with Progressive Candidates around the Country," *Verge*, March 8, 2018, theverge.com.

p. 144, **given enough eyeballs:** Ibid.

p. 144, **Open source development:** There are differences between free and open source software, in that open source software has been the subject of appropriation by technology capitalism and has lost much of the political nature of the free software movement. Here I use the term descriptively, rather than ideologically, for ease of reading.

p. 144, **Agile management focuses on:** Agile Manifesto, 2001, agilemanifesto.org.

pp. 144–5, **gateways to a more radical politics:** Wendy Liu, "Freedom Isn't Free," *Logic*, Failure edition, 2018, logicmag.io.

p. 146, **It did this in a variety of ways:** See, for example, *Caldera, Inc. v. Microsoft Corp.*, United States District Court for the District of Utah, 72 F. Supp. 2d 1295, November 3, 1999, law.justia.com.

p. 146, **What the [user] is supposed:** "Microsoft Emails Focus on DR-DOS Threat," CNet, January 2, 2012, cnet.com.

p. 146, **hopelessly dependent:** Moody, *Rebel Code*, 28.

pp. 146–7, **Labour produces for the rich:** Karl Marx, *Economic and Philosophical Manuscripts of 1844*, "Estranged Labour," marxists.org.

p. 149, **having bestowed upon him:** Essinger, *Ada's Algorithm*, 230.

p. 151, **unrelenting failure:** Eric Foner, *Tom Paine and Revolutionary America*, Oxford University Press, 2005, 3.

p. 151, **This episode left him:** Sean Monahan, "Reading Paine from the Left," *Jacobin*, March 2013, jacobinmag.com.

p. 151, **When I reflect on the horrid cruelties ... the power of Britain:** Humanus (pseudonym), "A Serious Thought," *Pennsylvania Journal*, October 18, 1775, bartleby.com.

p. 152, **the possibility of a republic:** Foner, *Tom Paine and Revolutionary America*, xxxi–ii.

p. 152, **one of the most popular and influential:** See ibid., xxvii. This is relative to population.

p. 152, **plan for raising a fund:** Amicus letter, June 1775, thomaspaine.org.

p. 152, **The earth, in its natural:** Thomas Paine, *Agrarian Justice*, 1999 digital ed., grundskyld.dk, 8.

pp. 152–3, **better understood as a universal capital grant:** Monahan, "Reading Paine from the Left."

p. 154, **The peer and the beggar ... to be preserved:** Thomas Paine, *Rights of Man, Part II*, thomaspaine.org.

p. 154, **poverty was a product of civilization:** Foner, *Tom Paine and Revolutionary America*, 251.

p. 155, **the present state:** Paine, *Agrarian Justice*, 15.

p. 155, **Those moments had a global impact:** See C. L. R. James, *The Black Jacobins*, Vintage Books, 1963.

p. 156, **its platform Free Basics:** Previously also known as Facebook Zero and internet.org. For a discussion of its earlier incarnation, see MacKinnon, *Consent of the Networked*, 123–4.

p. 156, **67 percent of India's population:** World Bank staff estimates based on the United Nations Population Division's World Urbanization Prospects, data.worldbank.org.

p. 157, **the biggest barriers:** Mark Zuckerberg, "Free Basics Protects Net Neutrality," *Times of India*, December 28, 2015.

p. 157, **the biggest reason why:** "Facebook Founder Mark Zuckerberg at IIT Delhi: As It Happened," *Wall Street Journal*, October 27, 2015, blogs.wsj.com.

p. 157, **Who could possibly be against this?:** Zuckerberg, "Free Basics Protects Net Neutrality."

p. 157, **a winning customer acquisition strategy:** Caroline O'Donovan and Sheera Frenkel, "Here's How Free Basics Is Actually Being Sold around the World," *BuzzFeed*, January 27, 2016, buzzfeed.com.

p. 157, **Rather than being a way:** Ibid.

p. 158, **For Facebook, it is a vital way:** Martínez, *Chaos Monkeys*, "Monetizing the Tumor."

p. 158, **Their pitch about access:** Rahul Bhatia, "The Inside Story of Facebook's Biggest Setback," *Guardian*, May 12, 2016, guardian.com.

p. 158, **Only about a third ... are Internet subscribers:** Telecom Regulatory Authority of India, *The Indian Telecom Services Performance Indicators October–December, 2017*, March 26, 2018, ii, trai.gov.in.

p. 158, **Infrastructure, more than individual:** Indeed, this is the view of the current Indian government. See ET Bureau, "First Base Is to Create a Digital Infrastructure: Ravi Shankar Prasad," *Economic Times*, February 1, 2016, economictimes.indiatimes.com.

p. 158, **Anti-colonialism has been economically catastrophic:** Abhimanyu Ghoshal, "Marc Andreesson Just Offended 1 Billion Indians with a Single Tweet," *Next Web*, February 2016, thenextweb.com.

p. 158, **Andreessen apologized:** Adi Narayan, "Andreessen Regrets India Tweets; Zuckerberg Laments Comments," *Bloomberg*, February 10, 2016, bloomberg.com.

p. 158, **In an analysis of the saga:** Rajat Agrawal, "Why India rejected

Facebook's 'Free' Version of the Internet," *Mashable*, February 9, 2016, mashable.com.

pp. 158–9, **Google announced a new project:** Sundar Pichai, "Bringing the Internet to More Indians—Starting with 10 Million Rail Passengers a Day," *Google Blog*, September 27, 2015, googleblog.blogspot.in.

p. 159, **Google got the monopoly:** Sreejith Panickar, "Why Facebook and Google Are in Digital India," *DailyO*, September 29, 2015, dailyo.in.

p. 159, **This project ultimately aims:** Nick Pinto, "Google Is Transforming NYC's Payphones into a 'Personalized Propaganda Engine'," *Village Voice*, July 6, 2016, villagevoice.com.

p. 161, **the term is about applying:** See, "What Is Net Neutrality?" ACLU, December 2017, aclu.org.

p. 161, **When professor Tim Wu:** Tim Wu, "Network Neutrality, Broadband Discrimination," *Journal on Telecommunications and High Tech Law*, 2 (2003), 145.

p. 161, **It is therefore important**: Ibid., 146.

p. 162, **the primary right:** Thomas Paine, "Dissertation on the First Principles of Government," 1795, press-pubs.uchicago.edu.

p. 162, **right to have rights:** Hannah Arendt, *The Origins of Totalitarianism*, Harcourt, 1967, 297–8.

p. 162, **cannot not want:** "Bonding in Difference, Interview with Alfred Arteaga," in Donna Landry and Gerald MacLean, eds., *The Spivak Reader: Selected Works of Gayatri Chakravorty Spivak*, Routledge, 1996, 28.

p. 162, **The business model demanded:** Tim Wu on *Fresh Air*, WHYY, National Public Radio, October 17, 2016.

p. 163, **Facebook has admitted:** Glenn Greenwald, "Facebook Says It Is Deleting Accounts at the Direction of the US and Israeli Governments," *Intercept*, December 20, 2017, theintercept.com; see also Nadim Nashif, "Facebook vs Palestine: Implicit Support for Oppression," Al Jazeera, April 10, 2017, aljazeera.com.

p. 163, **Google provides significant funding:** Kenneth Vogel, "Google Critic Ousted from Think Tank Funded by the Tech Giant," *New York Times*, August 20, 2017, nytimes.com.

p. 163, **It's clear to us:** Paul Mozur, "LinkedIn Said It Would Censor in China. Now That It Is, Some Users Are Unhappy," *Wall Street Journal*, June 4, 2014, blogs.wsj.com.

p. 163, **these companies will use:** Greenwald, "Facebook Says It Is Deleting Accounts."

p. 163, **Facebook's objective:** Sarah Peres, "Rigged," *TechCrunch*, November 9, 2016, techcrunch.com; see also Roger McNamee, "How to Fix Facebook—Before It Fixes Us," *Washington Monthly*, January–March 2018, washingtonmonthly.com.

p. 164, **It curates the news feed:** John Keegan, "Blue Feed, Red Feed," *Wall Street Journal*, May 18, 2016, graphics.wsj.com.

p. 164, **Emotional states can be transferred:** Adam D. I. Kramera, Jamie E. Guillory and Jeffrey T. Hancock, "Experimental Evidence of Massive-Scale Emotional Contagion through Social Networks," *PNAS* 11:24 (March 25, 2014), pnas.org.

p. 164, **There was immediate uproar:** Robert Booth, "Facebook Reveals News Feed Experiment to Control Emotions," *Guardian*, June 29, 2014, theguardian.com.

p. 164, **Commentators raised concerns:** Ibid.

p. 164, **Facebook's chief technology officer:** Mike Schroepfer, "Research at Facebook," October 2, 2014, newsroom.fb.com.

p. 165, **Concerns about the polarizing nature:** Nicholas Thompson and Fred Vogelstein, "Inside the Two Years That Shook Facebook—and the World," *Wired*, February 12, 2018, wired.com.

p. 165, **When it emerged in 2018:** Carole Cadwalladr and Emma Graham-Harrison, "Revealed: 50 million Facebook Profiles Harvested for Cambridge Analytica in Major Data Breach," *Guardian*, March 18, 2018, theguardian.com.

p. 165, **democratization of propaganda:** Bridle, *New Dark Age*, 234.

p. 165, **Social media platforms:** Maciej Cegłowski, "Build a Better Monster: Morality, Machine Learning, and Mass Surveillance," April 18, 2017, Emerging Technologies for the Enterprise, Philadelphia, idlewords.com.

p. 165, **the best way to attract:** Craig Silverman, Ellie Hall et al., "Hyperpartisan Facebook Pages Are Publishing False and Misleading Information at an Alarming Rate," *BuzzFeed*, October 20, 2016, buzzfeed.com.

p. 166, **84 percent of its revenue:** "Alphabet Inc. Today Announced Financial Results for the Quarter Ended September 30, 2018," Alphabet Investor Relations, October 25, 2018, abc.xyz/investor.

p. 166, **98 percent of takings:** "Facebook, Inc. Today Reported Financial Results for the Quarter Ended September 30, 2018," Facebook Investor Relations, October 30, 2018, investor.fb.com.

p. 166, **84 percent of the global digital ad market** Matthew Garrahan: "Google and Facebook Dominance Forecast to Rise," *Financial Times*, December 4, 2017, ft.com.

p. 166, **58 percent of online ad spending in America** Todd Spangler: "Amazon on Track to Be No. 3 in U.S. Digital Ad Revenue but Still Way Behind Google, Facebook," *Variety*, September 19, 2018, variety.com.

p. 166, **There is mounting pressure:** See, for example, Mark Zuckerberg's post, January 12, 2018, facebook.com/zuck/posts/10104413015393571.

p. 166, **the plot of a sci-fi novel:** McNamee, "How to Fix Facebook."

p. 166, **He is just one of many:** Julia Carrie Wong, "Former Facebook Executive: Social Media Is Ripping Society Apart," *Guardian*, December 12, 2017, theguardian.com; Olivia Solon, "Ex-Facebook President Sean

Parker: Site Made to Exploit Human 'Vulnerability'," *Guardian*, November 9, 2017, theguardian.com.

p. 167, **It has the capacity:** See Kristin Ross on the Communards' vision of the "universal republic," *Communal Luxury*, 23, 28.

p. 167, **Around 60 percent of Americans:** Jeffery Gottfried and Elisa Shearer, "News Use across Social Media Platforms 2016," Pew Research Center, May 26, 2016, journalism.org.

p. 167, **Google/Alphabet has a fully integrated:** See, for example, John Marshall, "A Serf on Google's Farm," *Talking Points' Memo*, September 1, 2017, talkingpointsmemo.com.

p. 167, **Amazon dominates the market:** Polly Mosendz, "Amazon Has Basically No Competition among Online Booksellers," *Atlantic*, May 30, 2014, theatlantic.com, Lina Ahmed, "Amazon's Antitrust Paradox," *Yale Law Journal* 126:710 (2017), 766.

p. 168, **concentration of economic power:** Ahmed, "Amazon's Antitrust Paradox."

p. 168, **a kingly prerogative:** 21 Cong. Rec. 2457 (1890), statement of Sen. Sherman, appliedantitrust.com.

p. 168, **The common warning:** Bridle, *New Dark Age*, 113.

p. 169, **utilities of the digital age:** Ahmed, "Amazon's Antitrust Paradox," 797.

p. 169, **social arrangements that permit:** Nancy Fraser, *Fortunes of Feminism*, Verso, 2013, "Reframing Justice in a Globalizing World."

p. 169, **When the rich plunder:** Thomas Paine, *Common Sense*, 1792, 36.

p. 169, **Tech workers all over the United States:** See techworkerscoalition.org. In India, see the Communist New Democratic Labor Front (NDLF) with their division for IT Employees (ITE): Michelle Chen, "Programmers in India Have Created the Country's First Tech-Sector Union," *Nation*, June 29, 2017, thenation.com; for the Forum for IT Employees, see facebook.com/ForumForITEmployees, and PTI, "IT Employees Set to Form Union as Layoffs Loom Large," *Economic Times*, May 23, 2017, economictimes.indiatimes.com.

p. 169, **Humanity faces actual existential threats:** Lizzie O'Shea, "Tech Capitalists Won't Fix the World's Problems—Their Unionised Workforce Might," *Guardian*, November 24, 2017, theguardian.com.

p. 170, **Only a handful of people:** Robert G. Ingersoll, "Thomas Paine," *North American Review* 155:429 (1892), 195.

p. 170, **For the most part … labor movement in America:** Foner, *Tom Paine and Revolutionary America*, 261, 264.

p. 171, **It commemorates the historic strike:** University of Melbourne, Historic Campus Tour, unimelb.edu.au.

p. 171, **All the main participants:** Peter Love, "Melbourne Celebrates the 150th Anniversary of Its Eight Hour Day," *Labour History*, vol. 91, November 2006, 193.

p. 172, **eventually come to be known:** Rosa Luxemburg, "What Are the Origins of May Day?" 1894, marxists.org.

p. 172, **The Cotton Mills Act:** Early Factory Legislation, Parliament of the United Kingdom, parliament.uk.

p. 172, **a general strike in Philadelphia:** Philip S. Foner, *May Day: A Short History of the International Workers' Holiday 1886–1986*, International Publishers, 1986, 8.

p. 172, **May Day marking:** Eric Hobsbawm, *Worlds of Labour: Further Studies in the History of Labour*, Weidenfeld and Nicolson, 1984, 76–7.

pp. 172–3, **Human endurance was ... members of a society:** Frank L. McVey, "The Social Effects of the Eight-Hour Day," *American Journal of Sociology* 8:4 (January 1903), 521.

p. 173, **The manner in which:** Friedrich Engels, *The Conditions of the Working Class in England*, Panther Edition, 2010, 107, marxists.org.

p. 173, **temporal and spatial confinements:** David Frayne, *The Refusal of Work: The Theory and Practice of Resistance to Work*, Zero Books, 2015, 72.

p. 173, **Wealth inequality is:** Organisation for Economic Cooperation and Development, *In It Together: Why Less Inequality Benefits All*, 2015, 15.

p. 173, **10 percent of households receive 28 percent of income:** Income inequality is also arguably able to be addressed through commonly accepted redistributive policies, such as taxation.

pp. 173–4, **10 percent of American households own 76 percent of wealth:** Christopher Ingraham, "If You Thought Income Inequality Was Bad, Get a Load of Wealth Inequality," *Washington Post*, May 21, 2015, washington post.com.

p. 174, **life expectancy has stalled:** Elizabeth Arias, "Changes in Life Expectancy by Race and Hispanic Origin in the United States 2013–2014," *NCHS Data Brief*, no. 244 (April 2016). See also Danny Dorling and Stuart Gietel-Basten, "Life Expectancy in Britain Has Fallen So Much that a Million Years of Life Could Disappear by 2058," *Conversation*, November 29, 2017, theconversation.com.

p. 174, **Death rates among white:** Gina Kolata, "Death Rates Rising for Middle-Aged White Americans, Study Finds," *New York Times*, November 2, 2015, nytimes.com.

p. 174, **factories that are often brutal:** See, for example, Charles Duhigg and David Barboza, "In China, Human Costs Are Built into an iPad," *New York Times*, January 25, 2012, nytimes.com.

p. 174, **Fatal explosions are common:** Jay Greene, "Riots, Suicides, and Other Issues in Foxconn's iPhone Factories," *CNet*, September 25, 2012, cnet.com.

p. 174, **Jack Qiu makes a compelling case:** Jack Qiu, *iSlavery*, University of Illinois Press, 2016, 61, 64–5.

p. 174, **Foxconn Suicide Express:** Ibid., 56–7.

p. 174, **tales of misery:** See, for example, Enough, enoughproject.org. See also Lee Simmons, "Rare-Earth Market," *Foreign Policy*, July 12, 2016, foreignpolicy.com.

p. 174, **Men, women and children mine:** Qiu, *iSlavery*, 22.

p. 174, **Planned obsolescence:** Jamie Campbell, "Company Breaks Open Apple Watch to Discover What It Says Is 'Planned Obsolescence'," *Independent*, April 25, 2015, independent.co.uk.

p. 176, **today's great paradox:** James Besson, *Learning by Doing*, Yale University Press, 2015, 21.

p. 176, **technology contributes to a loss:** Ibid., 21.

p. 176, **at risk is one of the most:** Natalie Kitroeff, "Robots Could Replace 1.7 Million American Truckers in the Next Decade," *LA Times*, September 25, 2016, latimes.com.

p. 176, **robots eating all the jobs:** Andreesson, "This Is Probably a Good Time to Say That I Don't Believe Robots Will Eat All the Jobs."

p. 177, **entrepreneurs will attempt to commodify:** See, for example, Tom Slee, *What's Yours Is Mine*, OR Books, 2015, 445–7, where the author discusses how commercial platforms that seek to scale community value systems, such as Airbnb, tend to undermine the sharing culture they were originally built upon.

p. 177, **Nearly 60 percent:** See "Characteristics of Minimum Wage Workers, 2015," BLS Reports, Report 1061, April 2016, bls.gov; and Susan J. Lambert, Peter J. Fugiel and Julia R. Henly, "Precarious Work Schedules among Early-Career Employees in the US: A National Snapshot," in *Employment Instability, Family Well-being, and Social Policy Network*, August 27, 2014, ssascholars.uchicago.edu, 6.

p. 177, **carry devices connected to the cloud:** Carrie Gleason and Susan J. Lambert, "Uncertainty by the Hour," static.opensocietyfoundations.org, 2.

p. 177, **Google Glass has been reincarnated:** Stephen Levy, "Google Glass 2.0 Is a Startling Second Act," *Wired*, August 18, 2017, wired.com.

p. 177, **Amazon is patenting:** Olivia Solon, "Amazon Patents Wristband That Tracks Warehouse Workers' Movements," *Guardian*, February 1, 2018, theguardian.com.

p. 177, **increasingly complex skill sets:** James Besson, "How Computer Automation Affects Occupations: Technology, Jobs, and Skills," Boston University School of Law, *Law and Economics Working Paper*, nos. 15–49, revised October 2016, 2.

pp. 177–8, **the nature and terms of work:** Natalie Kitroeff, "Robots Could Replace 1.7 Million American Truckers in the Next Decade," *LA Times*, September 25, 2016, latimes.com; see also Besson, "How Computer Automation Affects Occupations." As Besson concludes: "Automation tends to replace low-wage jobs with high-wage jobs."

p. 178, **The labor market is increasingly characterized:** See Srnicek and Williams, *Inventing the Future*, "Post-Work Imaginaries": "Across both North America and Western Europe, the labour market is now characterised by a predominance of workers in low-skilled, low-wage manual and service jobs (for example, fast-food, retail, transport, hospitality and warehouse workers), along with a smaller number of workers in high-skilled, high-wage, non-routine cognitive jobs."

p. 178, **half of all activities:** McKinsey Global Institute, *A Future That Works: Automation, Employment and Productivity*, 2017, mckinsey.com; see also Joe McKendrick, "Half of All U.S. Jobs Will Be Automated, but What Opportunities Will Be Created?" *ZDNet*, September 3, 2013, zdnet.com; see also Carl Benedikt Frey and Michael A. Osborne, "The Future of Employment: How Susceptible Are Jobs to Computerization?" September 17, 2013, 2, 44, oxfordmartin.ox.ac.uk.

p. 178, **six in ten** McKinsey Global Institute, *A Future That Works*, 4.

p. 178, **under- and unemployment is growing:** Carsey Institute, "Underemployment in Urban and Rural America, 2005–2012," Issue Brief no. 55, Fall 2012, scholars.unh.edu. "Underemployment (or involuntary part-time work) rates doubled during the second year of the recession, reaching roughly 6.5 percent in 2009." Figures indicate that the rate has only slightly dropped since then.

p. 178, **A group of people without work:** See Srnicek and Williams, *Inventing the Future*, "The Future Isn't Working."

p. 178, **their new electronic "colleagues":** Derek Thompson, "When Will Robots Take All the Jobs?" *Atlantic*, October 31, 2016, theatlantic.com.

p. 178, **jobless recoveries:** See Srnicek and Williams, *Inventing the Future*, "The Future Isn't Working."

p. 179, **work has also become:** Frayne, *The Refusal of Work*, 46.

p. 180, **a *political demand*:** Srnicek and Willams, *Inventing the Future*, "Post-Work Imaginaries."

p. 180, **Its impact will depend on policy choices:** Besson, *Learning by Doing*, 25.

p. 180, **For the first time since his creation:** John Maynard Keynes, "Economic Possibilities for Our Grandchildren," *Essays in Persuasion*, Norton, 1963.

p. 181, **The free development:** Karl Marx, *Grundrisse: Notebook VII—The Chapter on Capital*, marxists.org.

p. 181, **to hunt in the morning:** Karl Marx, *The German Ideology*, marxists.org.

p. 182, **More people are working:** Lydia Saad, "The '40-Hour' Workweek Is Actually Longer—by Seven Hours," August 29, 2014, *Gallup*, gallup.com. A third of Americans work on the weekend, with the figure shooting up to half if the person works multiple jobs. Bureau of Labour Statistics, American Time Use Survey, Chart 11, bls.gov/tus.

p. 182, **We are Wall Street:** Eric W. Dolan, "Chicago Traders Taunt 'Occupy Chicago' Protesters with 'We Are Wall Street' Leaflets," *Raw Story*, October 27, 2011, rawstory.com.

p. 182, **body-optimizing:** Analee Newitz, "Soylent 2.0: Rob Rhinehart's Cult of Foodlessness Kicks into High Gear," *Gizmodo*, August 4, 2015, gizmodo.com.

p. 182, **We have been trained too long:** Keynes, "Economic Possibilities for Our Grandchildren."

p. 183, **a concerted, open-minded discussion:** Frayne, *The Refusal of Work*, 220.

p. 183, **This also happened to be:** McVey, "The Social Effects of the Eight-Hour Day," 523.

p. 183, **At the unveiling:** "Eight Hours Memorial: The Monument Unveiled," *Melbourne Leader*, April 25, 1903, 23.

p. 183, **everyone should work less:** Frayne, *The Refusal of Work*, 38.

p. 183, **to give the worker time:** State Library of Victoria, "The Eight Hour Day," Ergo, ergo.slv.vic.gov.au, quoting *Report of the committee appointed by the Victorian Operative Masons' Society to enquire into the origin of the eight-hours' movement in Victoria, adopted by annual meeting, June 11, 1884*, Labor Call Print, 1912.

p. 183, **a public necessity:** McVey, "The Social Effects of the Eight-Hour Day," 526.

p. 184, **server shutdowns overnight:** Lauren Collins, "The French Counter-strike against Work Email," *New Yorker*, May 24, 2016, newyorker.com.

p. 184, **A similar law:** Private employees disconnecting from electronic communications during nonwork hours, Int 0726-2018, March 22, 2018, legistar.council.nyc.gov; Clayton Guse, "A New Bill Would Make It Illegal for NYC Bosses to Contact Employees after Work Hours," *Time Out*, March 22, 2018, timeout.com.

p. 184, **In the Netherlands:** See Srnicek and Williams, *Inventing the Future*, "Post-Work Imaginaries." See also Netherland Enterprise Agency, *Working Hours and Rest Times*, business.gov.nl.

p. 184, **Reducing the number pf working days:** See Alex Williams, "How Three-Day Weekends Can Help Save the World (and Us Too)," *Conversation*, April 26, 2016, theconversation.com.

p. 184, **The State of Utah:** David Rosnick and Mark Weisbrot, "Are Shorter Work Hours Good for the Environment? A Comparison of U.S. and European Energy Consumption," Center for Economic and Policy Research, December 2006, cepr.net.

p. 185, **In our society:** Dave Graeber, "On the Phenomenon of Bullshit Jobs: A Work Rant," *Strike Magazine*, no. 3 (August 2013), strikemag.org.

p. 186, **Technology implementation is challenging:** Besson, *Learning by Doing*, 60.

p. 186, **a combination of those bright ideas:** See Andrew Russell, "Hail the Maintainers," *Aeon Magazine*, April 7, 2016, aeon.co.

p. 186, **worker cooperatives are enterprises:** This is broadly the same definition as can be found in Gorm Winther and Richard Marens, "Participatory Democracy May Go a Long Way: Comparative Growth Performance of

Employee Ownership Firms in New York and Washington States," in *Economic and Industrial Democracy* (18), 1997. See also Elizabeth A. Hoffman, *Co-operative Workplace Dispute Resolution*, Gower Publishing, 2012, 9. The definition of worker cooperatives used by Hoffman has a distinctly political component, namely, as "organizations with equally shared worker ownership and egalitarian ideologies."

p. 186, **Republic Windows and Doors:** New Era Windows Cooperative, "Our Story," newerawindows.com.

p. 187, **it generates knowledge-sharing about work:** See John Pencavel, *Worker Participation*, Russell Sage Foundation, 2001, 14–16. See generally Winther and Mares, "Participatory Democracy May Go a Long Way." See also Henry Hansmann, *Ownership of Enterprise*, Harvard University Press, 1996, 77–8.

p. 187, **Structuring organizations in this way:** Winther and Mares, "Participatory Democracy May Go a Long Way," 399.

p. 187, **If we make a mistake ... more enjoyable:** Sarah van Gelder, "Three Years Ago, These Chicago Workers Took Over a Window Factory. Today, They're Thriving," *Yes Magazine*, October 9, 2015, yesmagazine.org.

p. 187, **knowledge-sharing networks:** Alex Rosenblat, "The Network Uber Drivers Built," *Fast Company*, January 9, 2018, fastcompany.com.

p. 187, **some experiments in organizing:** Michelle Chen, "Can Worker Co-ops Make the Tech Sector More Equitable?" *Nation*, December 21, 2017, nation.com.

p. 188, **destruction of the very values:** Slee, *What's Yours Is Mine*, 2015, chapter 9.

p. 188, **the Sharing Economy's complete neglect:** Ibid.

p. 190, **Others, like Adam Greenfield:** Greenfield, *Radical Technologies*, 205.

p. 190, **Writing in 1920:** Dennis Milner, *Higher Production by a Bonus on National Output*, George Allen and Unwin, 1920, basicincome.qut.edu.au.

p. 190, **in the UK:** Jon Stone, "British Parliament to Consider Motion on Universal Basic Income," *Independent*, January 20, 2016, independent.co.uk; and see Heather Stewart, "John McDonnell: Labour Taking a Close Look at Universal Basic Income," *Guardian*, June 6, 2016, theguardian.co.uk.

p. 190, **popular support:** Harriet Agerholm, "Universal Basic Income: Half of Britons Back Plan to Pay All UK Citizens Regardless of Employment," *Independent*, September 10, 2017, independent.co.uk.

p. 190, **Canada:** Daniel Tencer, "Hugh Segal, Champion of Basic Income, to Design Ontario's Pilot Project, *Huffington Post Canada*, June 24, 2016, huffingtonpost.ca; see also Hugh Segal, "A Universal Basic Income in Canada Is More Realistic than You Think," *Macleans*, April 20, 2018, macleans.ca.

p. 190, **A proposal was considered in Switzerland:** Debate with Swiss senator Andrea Caroni, "Is It Time for Universal Basic Income?" Al Jazeera, June 18, 2016, aljazeera.com.

p. 190, **An experimental program in Finland:** Jon Henley, "Finland to End Basic Income Trial after Two Years," *Guardian*, April 24, 2018, theguardian.com.

p. 190–1, **The decision has been controversial:** Ibid., see also Antti Jauhiainen and Joona-Hermanni Mäkinen, "Why Finland's Basic Income Experiment Isn't Working," *New York Times*, July 20, 2017, nytimes.com.

p. 191, **A form of it has existed:** Kate McFarland, "Brazil: Basic Income Startup Gives 'Lifetime Basic Incomes' to Villagers," *Basic Income Earth Network*, December 23, 2016, basicincome.org.

p. 191, **More recently, the elite:** Nancy Scola, "Facebook's Next Project: American Inequality," *Politico*, February 19, 2018, politico.com.

p. 191, **Libertarians have proposed:** Nathan Schneider, "Why the Tech Elite Is Getting behind Universal Basic Income," *Vice*, January 6, 2015, vice.com.

p. 191, **It serves as a facilitator:** Albert Wenger, "More on Basic Income and Robots," *Continuations*, July 7, 2014, continuations.com.

p. 191, **it has the potential to paper over:** Ben Tarnoff, "Tech Billionaires Got Rich off Us. Now They Want to Feed Us the Crumbs," *Guardian*, May 16, 2016, theguardian.com. See also Greenfield, *Radical Technologies*, 204. Greenfield argues that, in practice, a UBI will result "not in anything like total leisure and unlimited self-actualization, but in the further entrenchment of desperation and precarity."

p. 192, **It could be a technocratic lifeline:** Peter Frase, *Four Futures: Life After Capitalism*, Verso, 2016, "Communism: Equality and Abundance," "Socialism: Equality and Scarcity." See also Greenfield, *Radical Technologies*, 288 and 292. Here he argues that, in a possible future dubbed "Green Plenty," a UBI isn't necessary, because goods are free for the taking (a kind of communal luxury), whereas another possible future, which he labels "Stacks Plus," is an alternative society where the market and state are indistinguishable.

p. 192, **a marked improvement on current arrangements:** Patrick Butler, "Benefit Sanctions: They're Absurd and Don't Work Very Well, Experts Tell MPs," *Guardian*, January 8, 2015, theguardian.com.

p. 192, **repeatedly proven to be effective:** Charles Kenny, "Give Poor People Cash," *Atlantic*, September 25, 2015, theatlantic.com.

p. 193, **one alternative commonly suggested:** See Mark Paul, William Darity Jr. and Darrick Hamilton, "Why We Need a Federal Job Guarantee," *Jacobin*, April 2, 2017, jacobinmag.com.

p. 193, **Modeling suggests this**: Pavlina R. Tcherneva and L. Randall Wray, "Common Goals—Different Solutions: Can Basic Income and Job Guarantees Deliver Their Own Promises?" *Rutgers Journal of Law and Urban Policy* 2:1 (2005).

p. 199**, seek to enrol people:** Frayne, *The Refusal of Work*, 222.

p. 194, **enough to make leisure delightful:** Russell, "In Praise of Idleness."

p. 194, **universal basic services:** see, for example, UCL Institute for Global Prosperity, *Social Prosperity for the Future: A Proposal for Universal Basic Services* 2017.

p. 195, **Rather than focusing:** Ibid., 10.

p. 195, **we need a combination of these programs:** See Tcherneva and Wray, "Common Goals—Different Solutions."

p. 196, **O my body!:** Frantz Fanon, *Black Skin, White Masks*, Pluto, 1986, 232.

p. 196, **Born in Martinique:** Fanon, *Black Skin, White Masks*, 16; Immanuel Wallerstein, "Reading Fanon in the 21st Century," *New Left Review*, 57 (May–June 2009), 117.

p. 196, **He left his job ... the anti-colonial revolution:** Wallerstein, "Reading Fanon in the 21st Century," 117; Louis Ricardo Gordon, *What Fanon Said: A Philosophical Introduction to His Life and Thought*, Fordham University Press, 2015, 91–2.

p. 196, **form of skewed rationality:** Gordon, *What Fanon Said*, 2.

p. 196, **The loss of dignity:** Ibid., 98.

p. 197, **his other key works:** Wallerstein, "Reading Fanon in the 21st Century," 118.

p. 197, **a lengthy stint:** Gordon, *What Fanon Said*, 2.

p. 198, **to apply his thinking outside:** Ibid., 17. Gordon suggests that this is something literary and cultural critics do easily with white philosophers but less so with Fanon.

p. 198, **Fanon dedicated his life:** Ibid., 140.

p. 199, **I was responsible:** Fanon, *Black Skin, White Masks*, 112.

p. 200, **the subjective experience:** See Gordon, *What Fanon Said*, 48.

p. 200, **I am overdetermined:** Fanon, *Black Skin, White Masks*, 116.

p. 201, **Privacy gets depoliticized:** Gürses, Kundani and van Hoboken, "Crypto and Empire," 484.

p. 202 **The Algerian has brought:** Frantz Fanon, *Toward the African Revolution*, Grove Press, 1927, 102–3.

p. 202, **an extremely important ... the colonial configuration:** Frantz Fanon, *A Dying Colonialism*, Grove Press, 1965, 73.

p. 202, **to the only means of entering:** Ibid., 83.

p. 202, **to become a reverberating:** Ibid., 95.

pp. 202–3, **digital doppelgangers:** Harcourt, *Exposed*.

p. 203, **that Fanon was involved in:** See Daniel Thürer and Thomas Burri, *Max Planck Encyclopedia of Public International Law*, Oxford University Press, 2008, "Self-Determination."

p. 205, **Certain kinds of information:** Jack M. Balkin, "Information Fiduciaries and the First Amendment," *UC Davis Law Review* 49:1183, (2016), 1205.

p. 205, **fiduciaries owe a duty of care:** Ibid., 1207–8.

p. 205, **The Algerian combatant:** Fanon, *Toward the African Revolution*, 102–3.

p. 206, **One example of this idea:** See solid.mit.edu.

p. 206, **pods of data:** Daniel Weinberger, "Ways to Decentralize the Web," *Digital Trends*, December 27, 2016, digitaltrends.com.

p. 206, **Other examples of projects:** Chelsea Barabas, Neha Narula and Ethan Zuckerman, *Defending Internet Freedom through Decentralization: Back to the Future?* Center for Civic Media and the Digital Currency Initiative, MIT Media Lab, 2017, 35–6.

p. 206, **"re-decentralize" the web:** Liat Clark, "Tim Berners-Lee: We Need to Re-Decentralize the Web," *Wired*, February 6, 2014, wired.co.uk.

p. 206, **a robust and fertile community:** Barabas, Narula and Zuckerman, *Defending Internet Freedom through Decentralization*, 4.

p. 207, **Blockchain technology has great potential:** See generally Greenfield, *Radical Technologies*, 145ff; Robert Plant, "Can Blockchain Fix What Ails Electronic Medical Records?," *Wall Street Journal*, April 27, 2017, blogs.wsj.com.

p. 208, **There is no reciprocal respect:** Gordon, *What Fanon Said*, 91. Fanon himself expressed this in various ways, including in *The Wretched of the Earth*, page 2: "Decolonization is the encounter between two congenitally antagonistic forces that in fact owe their singularity to the kind of reification secreted and nurtured by the colonial situation."

p. 208, **You do not disorganize a society:** Fanon, *The Wretched of the Earth*, 3.

p. 208, **We do not expect:** Fanon, *Toward the African Revolution*, 105.

p. 209, **actually becoming brothers:** Ibid., 44. I have used the translation from Gordon, *What Fanon Said*, 91.

p. 209, **"revealed" the existence:** Jasper Hamill, "Could Google Maps Help End Poverty?" *Forbes*, January 28, 2014, forbes.com. Admittedly the software was provided through a private company, Google.

p. 209, **OpenStreetMap:** See wiki.osmfoundation.org.

p. 209, **The work of the platform:** Riva Richmond, "Digital Help for Haiti," *New York Times*, January 27, 2010, gadgetwise.blogs.nytimes.com.

p. 210, **employees can often reduce:** See Kathy L. Hudson and Karen Pollitz, "Undermining Genetic Privacy? Employee Wellness Programs and the Law," *New England Journal of Medicine*, May 24, 2017, nejm.org.

p. 210, **proposed legislative reforms:** See ibid.

p. 211, **Algerian society made:** Fanon, *A Dying Colonialism*, 84.

p. 211, **What do blacks want?:** Fanon, *Black Skin, White Masks*, 10. Note that this has been translated as "What does the black man want?" However, Gordon makes a convincing argument that the French is more universal, not specific to men. See Gordon, *What Fanon Said*, 21.

p. 211, **pull our preferences to the fringes:** Cegłowski, "Build a Better Monster."

p. 212, **Rather than being paid … and how to get it:** Richard Seymour, "On Forgetting Yourself," *Lenin's Tomb*, February 11, 2017, leninology.co.uk.

p. 212, **legalized crack:** Martínez, *Chaos Monkeys*, 276.

p. 212, **Natasha Schüll's conclusions:** Schüll, *Addiction by Design*, 20.

p. 213, **that the tool never ... a human world:** Fanon, *Black Skin, White Masks*, 231.
p. 213, **externally directed activity:** Gordon, *What Fanon Said*, 71–2.
p. 213, **honeypot for assholes:** Charlie Warzel, "'A Honeypot for Assholes': Inside Twitter's 10-Year Failure to Stop Harassment," *BuzzFeed*, August 11, 2016, buzzfeed.com.
p. 213, **Facebook's guidelines for content moderation:** Nick Hopkins and Olivia Solon, "Facebook Flooded with 'Sextortion' and Revenge Porn, Files Reveal," *Guardian*, May 22, 2017, theguardian.com.
p. 213, **we suck at dealing with abuse:** See Jenny Kutner, "Twitter CEO: 'We Suck at Dealing with Abuse and Trolls'," *Salon*, February 5, 2015, salon.com; see also "Why Twitter's 'Troll' Trouble Could Hurt Its IPO Prospects," CNBC, August 1, 2013, cnbc.com.
p. 213, **reportedly thwarted attempts:** Alex Hern, "Did Trolls Cost Twitter $3.5bn and Its Sale?" *Guardian*, October 18, 2016, theguardian.com.
p. 214, **Imperialism:** Fanon, *The Wretched of the Earth*, 181.
p. 214 **some examples of decay:** Fanon, *The Wretched of the Earth*, 181ff.
p. 214, **Honor, dignity and integrity ... a human being:** Ibid., 221.
p. 214, **Cade Diehm argues:** Cade Diehm, "On Weaponised Design," Our Data Our Selves, ourdataourselves.tacticaltech.org.
p. 215 **designers have offered:** Randi Lee Harper, "Putting Out the Twitter Trashfire," *ART+marketing*, February 13, 2016, artplusmarketing.com.
p. 215, **the bloodless genocide:** Fanon, *The Wretched of the Earth*, 238.
pp. 215–216, **They allow the rich:** The original quote is: "In its majestic equality, the law forbids rich and poor alike to sleep under bridges, beg in the streets and steal loaves of bread." Anatole France, *Le Lys Rouge* (The Red Lily) [1894].
p. 216, **The belief in European supremacy:** Irene Watson, "Settled and Unsettled Spaces: Are We Free to Roam?," in Aileen Morton-Robinson, ed., *Sovereign Subjects*, Allen and Unwin, 2007, 17.
p. 216, **The nation's *speech*:** Fanon, *A Dying Colonialism*, 3.
p. 216, **The same time:** Ibid., 179.
p. 217, **As part of a settlement:** Isaac Davison, "Whanganui River Given Legal Status of a Person under Unique Treaty of Waitangi Settlement," *New Zealand Herald*, March 15, 2017, nzherald.co.nz.
p. 217, **instead of the traditional model:** Eleanor Ainge Roy, "New Zealand River Granted Same Legal Rights as Human Being," *Guardian*, March 16, 2017, theguardian.com.
p. 217, **As well as financial redress:** Kennedy Warne, "Who Are the Tuhoe?" *New Zealand Geographic* 119 (January–February 2013), nzgeo.com.
p. 218, **The Colombian Supreme Court of Justice:** Community Environmental Legal Defense Fund, "Colombia Supreme Court Rules that Amazon Region Is 'Subject of Rights,'" April 6, 2018, intercontinentalcry.org.
p. 218, **They do not view the natural environment:** Roy, "New Zealand River

Granted Same Legal Rights."

p. 218, **one of their ancestors:** Warne, "Who Are the Tuhoe?"

p. 218, **entangled in this place:** Ibid.

p. 218, **companies and trusts have can be legal persons:** Jonathan Pearlman, "New Zealand River to Be Recognized as Living Entity after 170-year Legal Battle," *Telegraph*, March 15, 2017, telegraph.co.uk.

p. 218, **It's not that we've changed our worldview:** Davison, "Whanganui River Given Legal Status."

p. 218, **Indigenous communities:** First Nations people who were living on land that was subjected to colonization in New Zealand, Australia and Canada. How the descendants of the original inhabitants refer to themselves differs, of course, but for ease of reading I will use the term "Indigenous" throughout.

p. 218, **Colonial projects:** See Irene Watson, *Aboriginal Peoples, Colonialism and International Law: Raw Law*, Routledge, 2015, 5–6.

p. 219, **the longest lasting pan-continental stability:** Bruce Pasco, *Dark Emu*, Magabala Books, 2018, 185, 184.

p. 219, **there is overwhelming evidence:** Melissa Lucashenko, "The First Australian Democracy," *Meanjin Quarterly* 74:3 (2015), meanjin.com.au.

p. 219, **There are real questions:** See Watson, *Aboriginal Peoples, Colonialism and International Law*.

p. 219, ***Mana motuhake*:** Warne, "Who are the Tuhoe?"

p. 220, **conquest mentality … everything involved:** Taiaiake Alfred, "The Great Unlearning," February 28, 2017, taiaiake.net.

p. 220, **We live as a part:** Watson, *Aboriginal Peoples, Colonialism and International Law*, 15.

p. 221, **very few ever consider … growing up of humanity:** Irene Watson, "First Nations and the Colonial Project," *Inter Gentes* 1:1 (2016), 34, 39.

p. 221, **a human creation:** Taiaiake Alfred, "Sovereignty," in Joanne Barker, ed., *Sovereignty Matters: Locations of Contestation and Possibility in Indigenous Struggles for Self-Determination*, University of Nebraska Press, 2005, 46. Alfred argues that because "social and political institutions were designed and chartered by human beings means that people have the power and responsibility to change them." I note that while the Internet is a social creation, it is also a collective resource, meaning the metaphor is valid in some respects, even if the spiritual element is missing.

pp. 221–2, **Scholars of international law**: Ibid., 47.

p. 222, **fundamentally about inequality … late twentieth century:** Frase, *Four Futures*, Introduction: "Technology and Ecology as Apocalypse and Utopia."

p. 222–3, **Ownership was his obsession:** Warne, "Who Are the Tuhoe?"

p. 223, **"sovereignty" is an inappropriate political objective:** Alfred, "Sovereignty," in Barker, ed., *Sovereignty Matters*, 38.

p. 223, **Our goal should be:** Taiaiake Alfred, *Peace, Power, Righteousness: An*

Indigenous Manifesto, Oxford University Press, 1999, 21.

p. 223, **hopelessly clichéd images:** Trevor Paglen, "Smithsonian's Clarice Smith Distinguished Lecture," 2015, paglen.com.

p. 224, **a dreary vista:** See Zach Sokol, "Photographs of the Underwater Telecommunication Cables Tapped by the NSA," *Vice*, September 2015, vice.com.

p. 224, **perhaps most notoriously:** Andrew Blum, *Tubes*, Harper Collins, 2012, 5.

p. 224, **hard truths of geography:** Ibid., 9.

p. 224, **Undersea trunk cables:** Philip N. Howard, *Pax Technica*, Yale University Press, 2015, "The Empire of Connected Things."

p. 225, **The Merit network ... the national defense:** National Science Foundation, "About the National Science Foundation" nsf.gov; see also Merit, "A History of Excellence and Innovation," merit.edu; Eric M. Aupperle, "Merit—Who, What, and Why, Part One: The Early Years, 1964–1983," *Library Hi Tech* 16:1 (1998), eecs.umich.edu.

p. 225, **Throughout the early 1980s ... the Internet backbone:** Rajiv C. Shah and Jay P. Kesan, "The Privatization of the Internet's Backbone Network," *Journal of Broadcasting and Electronic Media* 51:1 (2007). For a visualization of this, see nsf.gov/news/special_reports.

p. 226, **backbone network traffic ... this exciting transition:** Aupperle, "Merit—Who, What, and Why."

p. 226, **This reflected ... expanded commercial Internet:** Shah and Kesan, "The Privatization of the Internet's Backbone Network."

p. 227, **relational philosophy ... bind future generations:** Watson, *Aboriginal Peoples, Colonialism and International Law*, 13–14.

p. 227, **the United States cannot match**: See *Akamai State of the Internet Report 2017*, 2017, akamai.com, 12.

p. 227, **speeds in rural areas:** Brian Whitacre, "Technology Is Improving—Why Is Rural Broadband Access Still a Problem?," *Conversation*, June 9, 2016 theconversation.com.

p. 227, **many Americans still lack:** *2016 Broadband Progress Report*, Federal Communication Commission, 2016, 2, fcc.gov.

p. 227, **made without any accountability:** Ben Tarnoff, "The Internet Should Be a Public Good," *Jacobin*, April 19, 2017, jacobinmag.com.

p. 228, **some academics shouldered this task:** Ramakrishnan Durairajan, Paul Barford, Joel Sommers and Walter Willinger, "InterTubes: A Study of the US Long-haul Fiber-optic Infrastructure," pages.cs.wisc.edu (published in SIGCOMM, "15 Proceedings of the 2015 ACM Conference on Special Interest Group on Data Communication, 565–78); see also Tom Simonite, "First Detailed Public Map of U.S. Internet Backbone Could Make It Stronger," *MIT Technology Review*, September 15, 2015, technology review.com.

p. 228, **The rapid accumulation:** Suarez-Villa, *Technocapitalism*, 23.

p. 228, **Some of the best aspects:** Jonathan Zittrain, *The Future of the Internet and How to Stop It*, Caravan Books, 2008, 27–8.

p. 229, **It reflected an ideological choice:** Tarnoff, "The Internet Should Be a Public Good."

p. 229 **love comparing the current internet:** Philip N. Howard, "Why the Internet Should Be a Public Resource," January 15, 2015, blog.yalebooks.com.

pp. 229–30, **the story of the greedy frog:** Watson, *Aboriginal Peoples, Colonialism and International Law*, 16.

p. 230, **collated data ... the national median:** Community Network Map, muninetworks.org.

p. 230, **making Chattanooga the first city:** Jason Koebler, "The City that Was Saved by the Internet," *Motherboard Vice*, October 27, 2016, motherboard.vice.com.

p. 230, **the city has established itself:** James O'Toole, "Chattanooga's Super-Fast Publicly Owned Internet," *CNN Tech*, May 20, 2014, money.cnn.com.

p. 230, **popular movement to reverse privatization:** Tarnoff, "The Internet Should Be a Public Good."

p. 231, **the primary goals:** Alfred, "Sovereignty," in Barker, ed., *Sovereignty Matters*, 46.

p. 231, **a wish to preserve:** Ibid., 45.

p. 231, **explicitly allow for difference:** Ibid., 46.

p. 232, **create a political ... and do things:** Ibid., 47–8.

p. 232, **These models operate:** Ben Schmidt, "Vector Space Models for the Digital Humanities," October 25, 2015, bookworm.benschmidt.org.

pp. 232–3, **a computer can learn ... Tokyo is to Japan:** Tolga Bolukbasi, Kai-Wei Chang, James Zou, Venkatesh Saligrama and Adam Kalai, "Man Is to Computer Programmer as Woman Is to Homemaker? Debiasing Word Embeddings," July 21, 2016, arxiv.org.

p. 233, **researchers have experimented:** "How Vector Space Mathematics Reveals the Hidden Sexism in Language," *MIT Technology Review*, July 27, 2016, technologyreview.com.

p. 233, **if the phrase:** Ibid.

p. 234, **Images of shopping:** Tom Simonite, "Machines Taught by Photos Learn a Sexist View of Women," *Wired*, August 21, 2017, wired.com.

p. 234, **has spoken about this:** Miriam Posner, "What's Next: The Radical, Unrealized Potential of Digital Humanities," keynote address at the Keystone Digital Humanities Conference, University of Pennsylvania, July 22, 2015, miriamposner.com.

p. 235, **our sense of the world:** Greenfield, *Radical Technologies*, 23.

p. 235, **renamed whole suburbs:** Jack Nicas, "As Google Maps Renames Neighborhoods, Residents Fume," *New York Times*, August 2, 2018, nytimes.com.

p. 235, **Ben Schmidt offers some insights ... but to understand it:** Ben

Schmidt, "Rejecting the Gender Binary: A Vector-Space Operation," October 30, 2015, bookworm.benschmidt.org; see also Claire Cain Miller, "Is the Professor Bossy or Brilliant? Much Depends on Gender," *New York Times*, February 6, 2016, nytimes.com; and "Gendered Language in Teaching Reviews—Interactive Chart," benschmidt.org.

p. 236, **while visiting Tahiti:** Greg Milner, *Pinpoint*, Granta, 2016, 7; Dan O'Sullivan, *In Search of Captain Cook*, I.B. Tauris, 2008, 147.

p. 236, **it is hard to overstate:** Milner, *Pinpoint*, 5.

p. 236, **Tupaia crafted a map:** O'Sullivan, *In Search of Captain Cook*, 148.

p. 237, **proud and austere:** Ibid., 150.

p. 237, **a cardinal sin:** Ibid.

p. 237, **the thrill in capturing:** Posner, "What's Next."

p. 238, **requirement for universal inclusion:** Alfred, *Peace, Power, Righteousness*, 91.

p. 239, **It began as a sale:** David Hackett Fischer, "Boston Common," in William E. Leuchtenburg, ed., *American Places: Encounters with History*, Oxford University Press, 2000, 128.

p. 239, **Boston Common belongs:** Samuel Barber, *Boston Common—A Diary of Notable Events, Incidences, and Neighboring Occurrences*, 2nd ed., Christopher Publishing House, 1914.

p. 240, **the conundrum and its possible solutions:** Elinor Ostrom, *Governing the Commons: The Evolution of Institutions for Collective Action*, Cambridge, 1990, 2–3.

p. 240, **conversion into private property:** James Boyle, "The Second Enclosure Movement and the Construction of the Public Domain," *Law and Contemporary Problems* 16:33 (2003), 34.

p. 241, **The town got together:** Barber, *Boston Common*, 22.

p. 241, **People often tested ... the town meeting:** Fischer, "Boston Common," in Leuchtenburg, ed., *American Places*, 128.

p. 241 **tragedy can be avoided:** Derek Wall, *Elinor Ostrom's Rules for Radicals*, Pluto Press, 2017, 34.

p. 241, **as an irrefutable argument:** David Harvey, "The Future of the Commons," *Radical History Review* 109 (Winter 2011), 101.

p. 241, **a revolution of the rich against the poor:** Karl Polanyi, *The Great Transformation: The Political and Economic Origins of Our Time*, Beacon Press, 2001, 35.

p. 241, **a long, slow, violent operation:** Peter Linebaugh and Marcus Rediker, *The Many-Headed Hydra*, Verso, 2012, 17.

p. 242, **Since the people:** Ibid.

p. 242, **significant riots:** Roger B. Manning, "Patterns of Violence in Early Tudor Enclosure Riots," *Albion: A Quarterly Journal Concerned with British Studies* 6:2 (Summer 1974), 120–33.

p. 242, **tradition of anti-authoritarian rioting:** Paul A. Gilje, *Rioting in America*, Indiana University Press, 1996, 12–13.

p. 242, **negroes and persons of vile condition … the working classes:** Barber, *Boston Common*, 66, 76–8.

p. 242, **Since the early … by skilled technicians:** See the Statute of Monopolies, 1623, legislation.gov.uk.

p. 243, **not as self-contained pieces … the design process:** Paul Mason, "The End of Capitalism Has Begun," *Guardian*, July 17, 2015, guardian.co.uk.

p. 243, **Through licensing:** see Joel Hruska, "General Motors, John Deere Want to Make Tinkering, Self-Repair Illegal," *Extreme Tech*, April 22, 2015, extremetech.com.

p. 243, **A man who built:** Tom Jackman, "E-waste Recycler Eric Lundgren Loses Appeal on Computer Restore Disks, Must Serve 15-Month Prison Term," *Washington Post*, April 29, 2018, washingtonpost.com.

p. 244, **a vote of no-confidence:** Boyle, "The Second Enclosure Movement," 40.

p. 244, **a second enclosure movement:** Ibid.

p. 245, **joining one person to another:** Fischer, "Boston Common," in Leuchtenburg, ed., *American Places*, 127.

p. 245, **We must be knit together:** John Winthrop, "City Upon a Hill," Sermon Aboard the Arbella, Heading en Route to Colonial America, 1630, worldhistoryproject.org.

p. 246, **Proposals to crack down:** Thomas Margoni, "Why the Incoming EU Copyright Law Will Undermine the Free Internet," *Conversation*, July 3, 2018, theconversation.com.

p. 246, **It is impossible:** Boyle, "The Second Enclosure Movement," 42.

p. 247, **floors full of lawyers:** Richard Anderson, "Pharmaceutical Industry Gets High on Fat Profits," BBC News, November 6, 2014, bbc.com.

p. 247–8, **Typically, when you evergreen:** Roger Collier, "Drug Patents: The Evergreening Problem," *CMAJ* 185:9 (June 11, 2013), E385, ncbi.nlm.nih.gov.

p. 248, **Tech companies Apple:** John Kell, "The 10 Most Profitable Companies of the Fortune 500," *Fortune*, June 11, 2015.

p. 248, **On average, pharmaceutical companies:** Anderson, "Pharmaceutical Industry Gets High on Fat Profits."

p. 249, **The Internet's inherent value:** MacKinnon, *Consent of the Networked*, 18.

p. 252, **There are many reasons:** See, for example, Eben Moglen, "Anarchism Triumphant: Free Software and the Death of Copyright," *First Monday* 4:8 (August 2, 1999), firstmonday.org.

p. 252, **One study found:** Yochai Benkler, *The Penguin and the Leviathan*, Random House, 2011, 182.

p. 253, **Well, the people:** George Johnson, "Once Again, a Man with a Mission," *New York Times Magazine*, November 25, 1990, nytimes.com.

p. 253, **there is research:** Benkler, *The Penguin and the Leviathan*, 13.

p. 253, **systematically, significantly and predictably:** Ibid., 14.

p. 254–5, **almost all the hackers:** There are some notable exceptions to this, including developers from Asian backgrounds. But these people still tended to be concentrated in the West. See generally Moody, *Rebel Code*.

p. 255, **They almost all hailed:** Söderberg, *Hacking Capitalism*, 28–9.

p. 255, **Benkler gives one example:** Benkler, *The Penguin and the Leviathan*, 153–8.

p. 256, **uneven geographies of participation:** Mark Graham, Ralph Straumann and Bernie Hogan, "Digital Divisions of Labor and Informational Magnetism: Mapping Participation in Wikipedia," *Annals of the Association of American Geographers* 105:6 (2015).

p. 256, **the democratizing potential:** Ibid.

p. 256, **the work of Elinor Ostrom ... a range of guidelines:** Ostrom, *Governing the Commons*; Elinor Ostrom, "Collective Action and the Evolution of Social Norms," *Journal of Economic Perspectives* 14:3 (2000), 137–58. See also Greenfield, *Radical Technologies*, 172–4.

p. 256, **In particular, her guidelines ... for resolving disputes:** Ostrom, *Governing the Commons*. See also Fernanda B. Viégas, Martin Wattenberg, and Matthew M. McKeon, "The Hidden Order of Wikipedia," hint.fm/papers.

p. 257, **They include the capacity:** Yochai Benkler, "Coase's Penguin, or, Linux and *The Nature of the Firm*," *Yale Law Review* 112 (2002), 369.

p. 258, **invitational quality:** Greenfield, *Radical Technologies*, 173–4.

p. 258, **Bought by Microsoft:** Tom Warren, "Microsoft Confirms It's Acquiring GitHub for $7.5 Billion," *Verge*, June 4, 2018, theverge.com.

p. 259, **More quantity makes:** Robert A. Burton, "How I Learned to Stop Worrying and Love AI," *New York Times*, September 21, 2015, opinionator.blogs.nytimes.com.

p. 259, **sent Garry into:** Rudy Chelminksi, "This Time It's Personal," *Wired*, October 1, 2001, wired.com.

p. 259, **If you wrap the Internet:** Moglen, "Anarchism Triumphant: Free Software and the Death of Copyright."

p. 260, **misattributes the sources:** Srnicek and Williams, *Inventing the Future*, "Conclusion."

p. 260, **the classified world:** Bridle, *New Dark Age*, 169.

Index